BIODIVERSITY

Challenges and Measures

THE EDITOR

Dr. Shashikanth H. Majagi did his M. Sc. and Ph. D. in Zoology from Gulbarga University, Kalaburagi, Karnataka. He started his carrier as Assistant professor in Government First Grade College, Chikkballapur and Government College (Autonomous) Kalaburagi. He has published 45 research papers and book articles. He has attended and presented 50 research papers in national/international conferences/symposia and 06 books to his credit. He is a reviewer/editorial board member for many journals of international acclaim. He is the life member for many academic bodies. He has given radio talks and Doordarshan programmes as well. Received prestigious Award for Research Paper (ARP) award from VGST, Government of Karnataka and IOSRD Life time award and Novel Research Academy Award of Excellence in Research. He has successfully organized two national Conference and Workshops. Successfully completed research projects funded by is UGC. Successfully Guided 02 Ph. D. students and 03 are working. He has worked as Special officer for NAAC, Department of Collegiate Education, Regional Joint Director Office, Kalaburagi. Currently he is teaching Environmental biology, Biodiversity and Limnology to Master degree Students. His research interests are on Limnology, Water quality studies and biodiversity. Currently working as Associate Professor and Chairman, Department of Zoology, Deputy Registrar (DPAR) and Nodal officer for Youth Red Cross.

BIODIVERSITY
Challenges and Measures

– Editor –

Shashikanth Majagi

Associate Professor and Chairman
Department of Studies in Zoology,
Vijaynagar Sri Krishnadevaraya University,
Ballari – 583105, Karnataka

2022

Daya Publishing House®
A Division of
Astral International Pvt. Ltd.
New Delhi – 110 002

ISBN: 9789354614095

Publisher's Note:

Every possible effort has been made to ensure that the information contained in this book is accurate at the time of going to press, and the publisher and author cannot accept responsibility for any errors or omissions, however caused. No responsibility for loss or damage occasioned to any person acting, or refraining from action, as a result of the material in this publication can be accepted by the editor, the publisher or the author. The Publisher is not associated with any product or vendor mentioned in the book. The contents of this work are intended to further general scientific research, understanding and discussion only. Readers should consult with a specialist where appropriate.

Every effort has been made to trace the owners of copyright material used in this book, if any. The author and the publisher will be grateful for any omission brought to their notice for Acknowledgment in the future editions of the book.

Published by : **Daya Publishing House**®
A Division of
Astral International Pvt. Ltd.
– ISO 9001:2015 Certified Company –
4736/23, Ansari Road, Darya Ganj
New Delhi-110 002
Ph. 011-43549197, 8130496929
e-mail: info@astralint.com
Website: www.astralint.com

— Dedicated to —

All those who have strived hard to protect and conserve the biodiversity

Acknowledgements

I wish to express my deep sense of gratitude to all the research paper contributors who were very generous for sending their original research papers.

I am thankful to my guide and mentor Dr. K Vijaykumar, Former Vice-Chancellor, Professor and Chairman, Department of Zoology, Gulbarga University, Kalaburagi for sharing his time and knowledge with me and for writing foreword to this book.

I Profusely thank Dr. K. Vijaykumar, Professor and Chairman, Department of Zoology, Gulbarga University, Kalaburagi and Dr. B. B. Hosetti Professor (Retd.) Kuvempu University, Shankarghatta, for reviewing the research papers.

My sincere thanks to my Parents Sri. Hanmantappa Majagi and Shakuntala Laxmibai, Wife Dr. Anita S. Majagi and daughters Miss. Sinchana and Miss. Saanidhya.

Shashikanth Majagi

Foreword

India is one of the world's most biologically and culturally diverse countries and has tremendous biodiversity, genetic as well as of species and ecosystems. It contains over 7 per cent of the world's biodiversity on 2.5 per cent of the Earth's surface. India being one of the mega biodiversity countries is also one of the most densely populated regions of the world. The country harbors four of the biodiversity hotspots out of the 36 global biodiversity hotspots: India displays significant biodiversity. One of seventeen megadiverse countries, it is home to 7.6 per cent of all mammalian, 12.6 per cent of all avian, 6.2 per cent of all reptilian, 4.4 per cent of all amphibian, 11.7 per cent of all fish, and 6.0 per cent of all flowering plant species.

The varied edaphic, climatic and topographic conditions and years of geological stability have resulted in a wide range of ecosystems and habitats such as forests, grasslands, wetlands, deserts, and coastal and marine ecosystems. Arid and semi-arid regions cover 38.8 per cent of India's total geographical area.

For India, conservation of biodiversity is crucial not only because it provides several goods and services necessary for human survival, but also because it is directly linked with providing livelihoods to and improving socio-economic conditions for millions of local people, thereby contributing to sustainable development and poverty alleviation. An example of a benefit derived from biodiversity in India is reflected by the forest sector, which is increasingly being looked upon as a major performer in poverty alleviation programmes. India's forests neutralize nearly 11 per cent of India's greenhouse gas emissions. Nearly 200 million people are dependent on forests for livelihood in India.

Rapid economic growth and limitations in integrating environmental concerns into development planning have put increasing pressure on biodiversity across

India, which is one of the globally recognized megadiverse countries rich in biodiversity. With only 2.4 percent of the earth's land area, India accounts for seven to eight percent of the world's recorded species. The world's largest social security scheme under the Mahatma Gandhi National Rural Employment Guarantee Act (2005) (MGNREGA) has created nearly five million green jobs in activities such as afforestation, water harvesting, soil conservation and land development.

The main threats to biodiversity include: habitat fragmentation, degradation and loss; over-exploitation of resources; shrinking genetic diversity; invasive alien species; declining forest resource base; climate change and desertification; impact of development projects; impact of pollution. In the backdrop of the varying socio-cultural milieu and often conflicting demands of various stakeholders, there is an urgent need for augmenting and accelerating the efforts for conservation and sustainable use of biodiversity, and for the fair and equitable sharing of benefits arising from the utilization of genetic resources.

The authors of the book chapters have put an effort to meet the needs of the book and audience. The present book has been well received in this direction, it covers the basic requirement of the knowledge for the biodiversity conservation, measures and challenges with specific case studies. The book will serve its immense use to the young researchers, scientists, teachers environmentalists and students of all disciplines

I congratulate the editor Dr. Shashikanth H. Majagi for bringing out this edited book at the right time comprehensively presenting various important areas like biodiversity, status, conservation, management and various biodiversity issues.

Dr. K. Vijaykumar

Former Vice-Chancellor,
Professor and Chairman
Department of Zoology
Gulbarga University,
Kalaburagi, Karnataka

Preface

India has four biodiversity hotspots, including the Himalayas, the Western Ghats, the Indo-Burma region and the Sundaland, that hosts numerous endemic species. Around 24.46 per cent of the country's geographical area covered by forests and trees. India has lost 90 per cent of the area under its four biodiversity hotspots, with the Indo-Burma hotspot taking the worst hit, a report released by Centre for Science and Environment. The report, titled State of India's Environment in Figures 2021, shows at least 25 species have also gone extinct in these hotspots. The International Union for Conservation of Nature's (IUCN) Red List monitors 1,212 animal species in India and have tagged 12 per cent as endangered.

Despite lot of actions taken up by the government and other bodies/agencies to conserve and protect the biodiversity but yet we are not able to achieve the desired target. Hence, still more actions and strict implementations needs to be taken for the conservation. The north Indian cities are highly polluted and out of the most polluted cities of the world Indian cities are 13 out of 20. This is alarming situation for India as well as world too and the water quality situation is also not different. We need to address the all issues to achieve the SDG goals.

With this set of back drop, this book, which contains papers from various topics related to biodiversity conservation and challenges. It provides an updated information of many of the principal issues in conservation of species and its resource. It also documents a new alliance between scientific, governmental, and commercial forces—one that can be expected to reshape the international conservation movement for decades to come. The 16 chapters of book of wide varieties offers an overall view of this biological diversity and carries the urgent warning that we are rapidly altering and destroying the environments that have fostered the diversity of life forms for more than a billion years.

Shashikanth Majagi

Contents

Chapter 1

Global Climate Change: Consequences and Future Strategies

Shabir A. Wani, Shabir A. Bhat, A.K. Tantray, Z.I. Buhroo, Shahina Nagoo and Masarat Bashir

College of Temperate Sericulture Mirgund, Sher-e-Kashmir University of Agricultural Sciences and Technology of Kashmir, J&K, India

Global warming is the observed and projected increase in the average temperature of earth's atmosphere and oceans. The earth's average temperature has rise by 0.6°C in the last 140 years. Fine dust or aerosol particles floating high in the atmosphere can alter the heat balance of the earth by reflecting away the incoming solar heat, thereby making the surface cooler and by reflecting the escaping heat down towards the surface making it warmer (Mackenzie, 1997). The balance of these two effects depends on the nature of the dust, season and time of the day for example a dust layer may tend to make the summer cooler and the winter warmer. The atmosphere is substantially influenced by the action of UV-radiation in ozone and water vapour (Crutzen, 1993). The main causes of global warming are greenhouse gases, power plants, pollution, population, deforestation and transport.

Global warming has effects and consequences on all walks of life. The consequences of global warming can be seen in the atmospheric weather, local climate change, glacier retreat and disappearance, oceans, seal level rise, acidification, forest fires, ozone depletion, agriculture, water scarcity as well as the health of individuals (Roy D-Silva, 2007). An increase of 2°C in temperature could decrease the rice yield by about 0.75 t/ha and 0.5°C increase in winter temperature reduce wheat yield 0.45 t/ha (Sinha and Swaminathan, 1991). Shreshtha et al. (1998) showed that about 7.4, 8.7 and 9.8 per cent of total cumulative CO_2, SO_2 and NO_2 emission respectively could be avoided between 1997 and 2015 by using efficient appliances. Improved training and general education of populations dependent on agriculture. Agriculture research to develop new crop varieties. Identification of the present vulnerabilities of agricultural *systems. Food programmes and other social security programmes to provide insurance against supply changes. Transportation, distribution and market integration to provide the infrastructure to supply food during crop short falls. It is imperative that the developed countries and the rapidly developing countries formulate strategies to curb greenhouse gas emissions. Countries on the fast tract of economic growth should also look at adopting new energy-saving technologies and planting of more trees. The emphasis should also be laid on increasing the use of renewable energy sources like solar and wind. It is high time for leading emitters of CO_2 to formulate national programmes to address climate change.*

Keywords: *Climate, Greenhouse gases, Global and Strategies etc.*

Introduction

The present day atmosphere has evolved from a mixture of gases released from the interior of the earth. The main gases released by volcanic activity are water vapour (64 per cent, w/w) CO_2 (24 per cent), SO_2 (10 per cent) and nitrogen (1.5 per cent). However, this mixture escaping from volcanoes over billion of years and acted upon by living organisms has produced an atmosphere which is mainly constituted by nitrogen and oxygen and a trace of CO_2 and other gases. In recent years environmental problem has developed to the global scale. The phenomena, such as ozone layer depletion, CO_2 increase, weather change, acid rain and oceanic pollution are treated as the problems of global cycle (Fuwa, 1992). On the other hand, fine dust or aerosol particles floating high in the atmosphere can alter the heat balance of the earth by reflecting the escaping heat down towards the surface making it warmer (Mackenzie, 1997). The balance of these two effects depends on the nature of the dust, season and time of the day, for example, a dust layer may tend to make the summer cooler and the winter warmer.

Greenhouse gases keep our planet warm. They exist naturally in the atmosphere, heating it by trapping energy that has originally come from the sun. This process is called the natural greenhouse effect. Without the greenhouse gases the surface of the earth would be as cold as the surface of the moon, about -18°C. In fact the average surface temperature of the earth is about 15°C.

In the last 100 years the atmosphere has warmed up by over half a degree Celsius (°C). Also during this time humans have been releasing extra greenhouse gases which are the result of burning fossil fuels (like coal, oil and gas). Greenhouse gases are also released by the exhaust of motor vehicle and by the cutting down of forests.

If the earth continues to warm as predicted, the temperature of the earth's surface may be 3°C warmer by the end of the 21st century. This rapid change in temperature would have many impacts on environment and on society. Global warming will affect rainfall, sea level and the number and strength of extreme weather events such as storms and floods. Human health, agriculture and water supplies will all be affected.

Climate change and agriculture are interrelated processes, both of which take place on a global scale. Global warming is projected to have a significant impacts on conditions affecting agriculture, including temperature, carbon-dioxide, glacial run-off, precipitation and the interaction of these elements. These conditions determine the carrying capacity on the biosphere to produce enough food for the human population and domesticated animals. The overall effect of climate change on agriculture will depend on the balance of animals. The overall effect of climate change on agriculture will depend on the balance of these effects. Assessment of the effects of global climate changes on agriculture might help to properly anticipate and adapt farming to maximize agricultural production.

As the same time, agriculture has been shown to produce significant effects on climate change, primarily through the production and release of greenhouse gases such as carbon-dioxide, methane, and nitrous oxide, but also by altering the earth's land cover, which can change its ability to absorb or reflect heat and light, thus contributing to radiative forcing. Landuse change such as deforestation and desertification, together with use of fossil fuels, are the major anthropogenic sources of carbon-dioxide; agriculture itself is the major contributor to increasing methane and nitrous oxide concentrations in earth's atmosphere. Global warming is the observed and projected increase in the average temperature of earth's atmosphere and oceans. The earth's average temperature has rise by 0.6°C in the last 140 years. Global warming is an increase in global average surface temperature due to natural or anthropogenic climate change (Wikipedia.org).

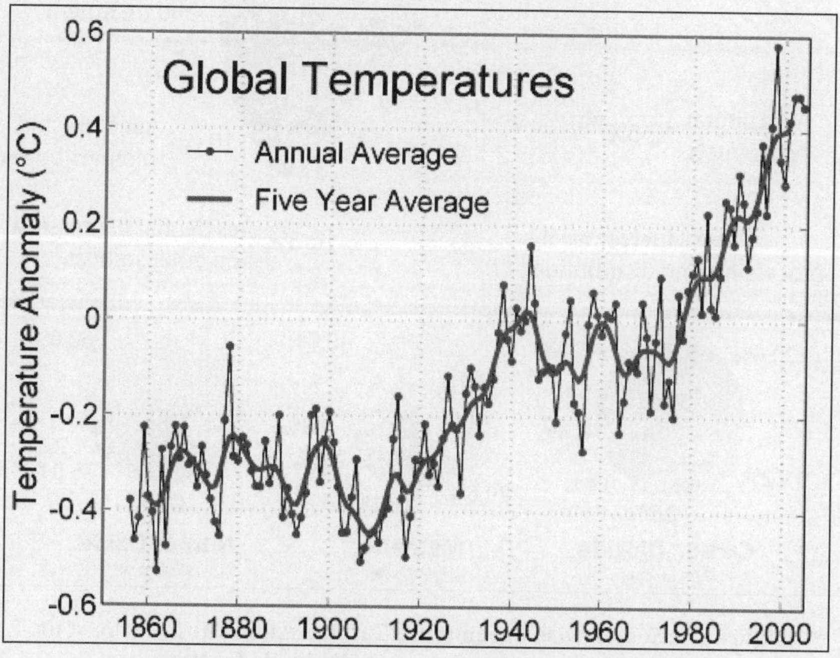

Figure 1.1: Definition for Global Warming: Temperature Increase in the Last 140 Years.

It is clear from the above graph that the global temperature shows the increasing trend in the temperature all over the world.

Greenhouse Gases

- ☆ Greenhouse gases are the main causes of the global warming.
- ☆ Greenhouse gases like carbon-dioxide, methane and nitrous oxide are playing hazards in the present time.
- ☆ These greenhouse gases trap heat in earth's atmosphere and thus results in increasing the temperature of earth.
- ☆ The excessive emission of these gases is the major cause of global warming.

(Source : www.globalwarming.org.in)

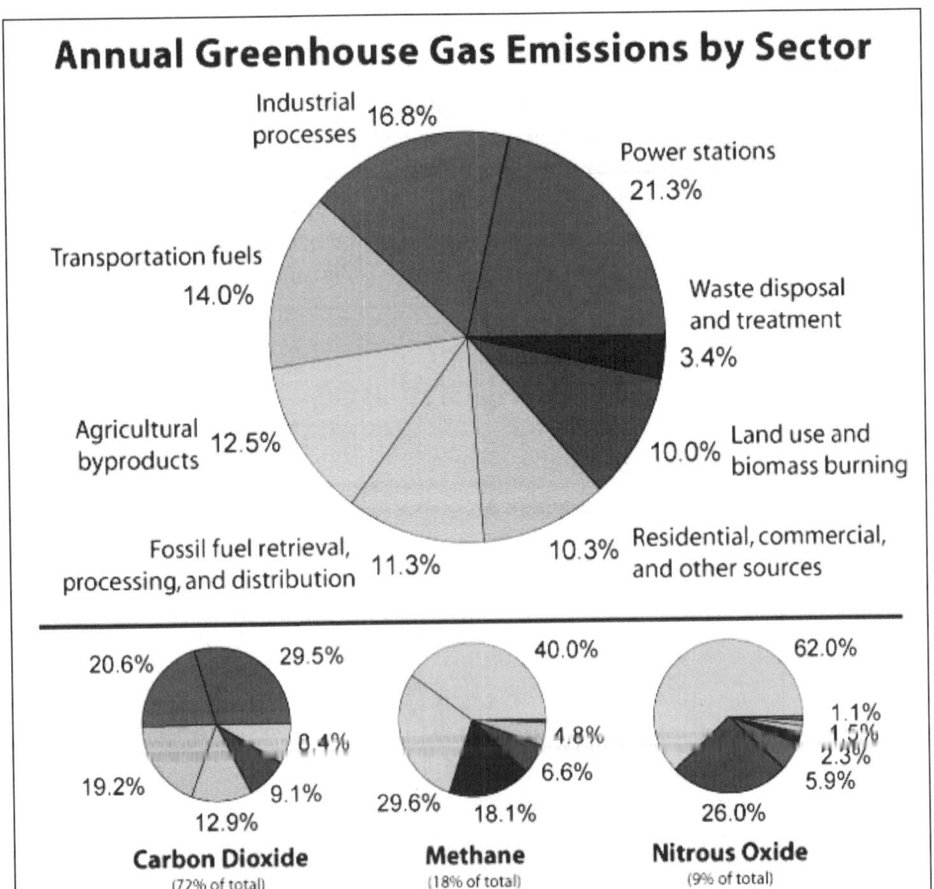

Figure 1.2: Sector-wise Greenhouse Gas Emission. The lower part of the picture shows the sources individually for the gases carbon-dioxide, methane and nitrous oxide, respectively.
(*Source*: http://en.wikipedia.org/wiki/Image:Greenhouse_Gas_by_Sector.png)

The above figure reveals that the power stations are the major source of greenhouse gas emission which is followed by industrial processes. Waste disposal and treatment are the least greenhouse emission sector. The increase of greenhouse gas concentration mainly carbon-dioxide led to a substantial warming of the earth and the sea called global warming.

Power Plants

The major source of carbon-dioxide is the power plants. Power plants emit large amounts of carbon-dioxide produced from burning of fossil fuels for the purpose of electricity generation. Coal is the major fuel that is burnt in these power plants. Coal produces around 1.7 times as much carbon-dioxide per unit of energy when flamed as does natural gas and 1.25 times as much as oil.

(Source : www.globalwarming.org.in)

Pollution

Thick haze and smoke, originating from burning biomass in northwestern India and air pollution from large industrial cities in northern India, often concentrate inside the Ganges basin. Dust and black carbon which are blown towards higher altitudes by winds at the southern faces of the Himalayas, can absorb shortwave radiation and heat the air over the Tibetan plateau. The net atmospheric heating due to aerosol absorption causes the air to warm and convert upwards, increasing the concentration of moisture in the mid-troposphere and providing positive feedback that stimulates further heating of aerosols (Lau, 2006).

Global Warming and Population

There has been a recent interest in associating climate change/global warming with "over population" and countries such as China and India have to do more to help contain global warming. Yet rich countries have a lot to do themselves. The emission from rich countries that accumulated in the atmosphere for so long to trigger climate change (Mishra *et al.*, 2007).

Deforestation

Deforestation causes 25 per cent of all carbon-dioxide release entering the atmosphere, by the cutting and burning of about 34 million acres of trees each year. Cutting of trees is leading to greater concentration of carbon-dioxide in the atmosphere. Greater urbanization, requirement of land for factories and buildings, requirement of timber are also reasons that are leading to deforestation which in turn is leading to global warming (www.globalwarming.org.in).

Transport

Another major source of carbon-dioxide in the atmosphere is the emission from cars and other vehicles. About 20 per cent of carbon-dioxide emitted in the atmosphere comes from burning of gasoline in the engine of the vehicles. It is always better to use vehicles designed for city driving on the city roads (www.globalwarming.org.in).

Major CO_2 Emitting Countries

Country	*CO_2 Emissions (in billion tones)*
USA	5.9
CHINA	4.7
RUSSIA	1.7
JAPAN	1.3
INDIA	1.1

Source: Guardian, 2007.

Effects of Global Warming

- ☆ Economy
- ☆ Environment
- ☆ Global dimining
- ☆ Extreme Weather
- ☆ Local Climate change
- ☆ Glacier retreat and disappearance
- ☆ Retreat of the Helheim, greenland
- ☆ Oceans
- ☆ Sea level rise
- ☆ Temperature rise
- ☆ Acidification
- ☆ Methane released from hydrates
- ☆ Forest fires
- ☆ Sulphur aerosols
- ☆ Ozone depletion

Effect on the Economy

The Indira Gandhi Institute of Development Research has reported that climate-related factors could cause India's Gross Domestic Products (GDP) to decline by upto 9 per cent. Around seven million people are projected to be displaced due to, among other factors, submersion of parts of Mumbai and Chennai, if global temperature's were to rise by a mere 2°C (Sethi *et al.*, 2007). Villagers in India's North Eastern State of Meghalaya are also concerned that rising sea level will submerge neighbouring low-lying Bangladesh resulting in an influx of refugees into Meghalaya. If server climate changes occur, Bangladesh will lose land along the coast line (Ahmad *et al.*, 2006).

Effect on Environment

Since the beginning of the industrial revolution in the second half of the 18th century, the levels of pollutant increased in the atmosphere by as much as 33 per cent. Flooding, coastal erosion, droughts, salt water intrusion into soils, unusual high levels of tropical storm, cyclones, huge loads of rain are among the effects seen in various parts of the world during the past years which have been attributed to global temperature rise. The rising sea levels cause flooding in many parts of the world. Over the past 100 years, the sea level has risen by almost 1 meter as a result of warming of our planet (Anup *et al.*, 2006).

Extreme Weather

Global warming may be responsible on pastfor some trends in natural disasters such as extreme weather. There will be increased intense tropical cyclone activity. There will be increased incidences of extreme high sea level (Stefan *et al.*, 2007).

Local Climate Change

The first recorded south Atlantic hurricane "Catarina" which hits Brazil in March, 2004. In the northern hemisphere, the southern part of the Arctic region has experienced a temperature rise of 1 to 3°C over the last 50 years. Canada Alaska and Russia are experiencing initial melting of permafrost. This may disrupt ecosystems and by increasing bacterial activity in the soil lead to these areas becoming carbon sources instead of carbon sinks (Vladimir and Romanovsky, 2007).

Glacier Retreat and Disappearance

Excluding the ice caps and ice sheets of the Arctic and Antarctic, the total surface area of glaciers worldwide has decreased 50 per cent since the end of the 19th century. Currently glacier retreat rates and mass balance losses have been increasing in the Andes, Alps, Pyrenels, Himalayas, Rocky mountains and North Cascades. The loss of glaciers not only directly cause landslides, flash floods and glacial lake overflow but also increases annual variation in water flows in rivers (Mauri and Pelto, 2007).

Oceans

The oceans serve as a sink for carbon-dioxide, taking up much that would otherwise remain in the atmosphere but increase levels of CO_2 have led to ocean acidification. As the temperature of oceans increases, they become less able to absorb excess CO_2. Global warming is projected to have a number of effects which include rising sea level due to thermal expansion and melting of glaciers and ice sheets (Christian *et al.*, 2003).

Sea Level Rise

With increasing average global temperature, the water in the oceans expands in volume and additional water enters than which has previously been locked up on land in glaciers, *e.g.* Greenland and Antarctic ice sheets. For most glaciers worldwide, an average volume loss of 60 per cent until 2050 is predicted. The total ice melting rate over Greenland is 2.39±23 cubic km per year. Instead of a global 5 meter sea level rise. The United States, parts of Canada and the Indian Ocean would experience upto 6.5 meters of sea level rise (Christian *et al.*, 2003).

Temperature Rise

From 1961 to 2003, the global ocean temperature has risen by 0.10°C from the surface to a depth of 700 m. The temperature of the Antarctic southern ocean rose by 0.17°C (0.31°F) between the 1950s and the 1980s nearly twice the rate for the world's oceans as a whole. Having effects on ecosystems by melting sea ice, warming reduces the ocean's ability to absorb CO_2 (Bindoff *et al.*, 2007).

Acidification

Ocean acidification is an effect of rising concentration of CO_2 in the atmosphere. Oceans currently absorb about one tonne of CO_2 per person per year. It is estimated that the oceans have absorbed around half of all CO_2 generated by human activities (Sabine *et al.*, 2004).

Methane Release from Hydrates

Methane clathrate also called methane hydrate is a form of water ice that contains a large amount of methane within its crystal structure. The sudden release of large amounts of natural gas from methane clathrate deposits in a runaway greenhouse effect has been hypothesized as a cause of post and possibly further climate changes. The release of this trapped methane is a potential major outcome of a rise in temperature (Connor, 2008).

Forest Fires

The Inter-Governmental Panel on Climate Change (IPCC) Fourth Assessment Report predicts that many mid-latitude regions such as Mediterranean Europe, will experience decreased rainfall and an increased risk of drought, which in turn would allow forest fires to occur in large scale. This releases more stored carbon into the atmosphere than the carbon cycle can naturally re absorb, as well as reducing the overall forest area on the planet, creating a positive feedback loop. Part of the feedback loop is more rapid growth of replacement forests and a northward migration of forests as northern latitude became more suitable climates for sustaining forests (USEPA, 2007).

Effect on Sulfur Aerosols

Sulfur aerosols, especially stratospheric sulfur aerosols have a significant effect on climate. One source of such aerosols is the sulfur cycle where plankton release gases which eventually becomes oxidised to sulfur dioxide in the atmosphere. Disruption to the oceans as a result of ocean acidification or disruptions to the thermohaline circulation may result in disruption of the sulfur cycle, thus reducing its cooling effect on the planet through the creation of stratospheric sulfur aerosols (Lawrence *et al.,* 2008).

Ozone Depletion

There are two types of stratospheric ozone depletion, the conventional type and the ozone hole type. It has been reported that the depletion of ozone is due to technology progress in rocketing, supersonic aircraft flight, refrigeration, nitrogenous fertilizers and others. An abundance of ozone near the earth's surface has resulted in damaging effects on human health, agriculture crops, ornamental plants, forests and materials. It has become clear that a decrease in emissions of nitrogen oxides may be needed in areas where ozone concentrations are high (Thukral and Virk, 2000).

Effect on Agriculture

Sinha and Swaminathan (1991) observed that an increase of 2°C in temperature could decrease the rice yield by about 0.75 tons/ha in the high yield areas and 0.5°C increase in winter temperature would reduce wheat yield by 0.45tonns/ha. Rao and Shina (1994) reported that wheat yield could decrease between 28-68 per cent without considering the carbon-dioxide fertilization effects. Saseendran *et al.* (2000) reported that for everyone degree rise in temperature the decline in rice yield would be about 6 per cent.

Soil Process

The potential for soils to support agriculture and distribution of land use will be influenced by changes in soil water balance. Increase in soil water deficits *i.e.* dry soils become drier, therefore increased need for irrigation. Could improve soil workability in wetter regions and diminish poaching and erosion risk (Zarin, 2007).

Crops

Horticultural crops are more susceptible to changing conditions than arable crops. Field vegetables will be particularly affected by temperature changes. Water deficits will directly affect fruit and vegetable production (Zarin, 2007).

Effect on Quality

For rice, the amylose content of the grain a major determinant of cooking quality is increased under elevated CO_2 (Conroy *et al.,* 1994). Cooked rice grain from plants grown in high CO_2 environments would be firmer than that from today's plants. However, concentration of iron and zinc which are important for human nutrition would be lower. The protein content of the grain decreases under combined increase of temperature and CO_2 (Ziska *et al.,* 1997).

Effect on Growing Period

An increase in temperature will speed up development. In case of an annual crops, the duration between sowing and harvest will shorten (*e.g.* the duration in order to harvest corn could shorten between one and four weeks). The shortening of such a cycle could have an adverse effect on productivity because senescence would occur sooner (Royal Society London, 2005).

Shortage in Grain Production

Crops such as sunflowers can be affected by severe drought conditions in Australia. Between 1996 and 2003, grain production has stabilized slightly over 1800 million of tons. In 2000-2003 grain stocks have been dropping, resulting in a global grain harvest that was short of consumption by 93 million tons in 2003 (Lobell *et al.*, 2007).

Effect on Pests and Diseases

Global warming would cause an increase in rainfall in some areas, which would lead to an increase of atmospheric humidity and the duration of wet seasons. Combined with higher temperature, these could favour the development of fungal diseases. Similarly with highly temperatures and humidity, there could be an increased pressure from insects and disease vectors (Alexey *et al.*, 2007).

Improving Adaptive Capability of Agriculture

Improving training and general education of population dependent on agriculture. Agriculture research to develop new crop varieties. Identification of the present vulnerabilities of agriculture systems. Food programmes and other social security programs to provide insurance against supply changes. Transportation, distribution and market integration to provide the infrastructure to supply food during crop short falls (Zarin, 2007).

Water Scarcity

Sea level rise is projected to increase salt-water intrusion into groundwater in some regions, affecting drinking water and agriculture in coastal zones. Increased evaporation will reduce the effectiveness of reservoirs. Increased extreme weather means more water falls on hardened ground unable to absorb it, leading to flash floods instead of a replenishment of soil moisture or groundwater. In some areas shrinking glaciers threaten the water supply (Hitz *et al.*, 2007).

Health

Climate change currently attributes to the burden of disease and premature deaths. The negative health impacts of climate change will outweigh the benefits, especially in developing countries. Negative health impact include increased malnutrition, increased deaths, disease and injury due to heat waves, floods, storm, fires and droughts and increased frequency of cardio-respiratory diseases. According to a 2009 study by UCL academics, climate change and global warming pose the biggest threat to human health in the 21st century (Lister, 2009).

Spread of Disease

Global warming may extend the favourable zones for vectors conveying infectious disease such as dengue, fever, west Nile virus and malaria. The World Health Organization (WHO) says global warming could lead to a major increase in insect-borne diseases in Britain and Europe as North Europe becomes warmers. In the United States, Malaria has been endemic in as much as 36 states until the 1940s.

Solutions to Global Warming

When you buy your car, look for the one with the best fuel economy in it class. Each gallon of gas you use is responsible for 25 pounds of heat-trapping gases in the atmosphere. Better gas mileage not only reduces global warming, but will also save your thousands of dollars at the pump over the life of the vehicle.

Choose Clean Power

More than half the electricity in the United States comes from polluting coal-fired power plants. Power plants are the single largest source of heat trapping gas. None of us can live without electricity, but in some states, you can switch to electricity companies that provide 50-100 per cent renewable energy (Ponce, 2009).

Look for Energy Star

When it comes time to replace appliances, look for the energy star label on new appliances (refrigerators, freezers, furnaces, air conditioners and water heaters use the most energy. Household energy savings really can make a difference. If each household in the United States replaced its existing appliances with the most efficient models available, we would save #15 billion in energy costs and eliminate 175 million tonnes of heat-trapping gases (Ponce, 2009).

Unplug a Freezer

One of the quickest ways to reduce your global warming impact is to unplug the extra refrigerator or freezer you rarely use. This can reduce the typical family's carbon-dioxide emissions by nearly 10 per cent (Ponce, 2009).

Plant a Tree

Get a group in your neighbourhood together and contact your local arborist or urban forester about planting trees on private property and public land. In addition to storing carbon, trees planted in and around urban areas and residences can reduce energy bills and fossil fuel use (Ponce, 2009).

Let Policy Makers know you are Concerned about Global Warming

Our elected officials and business leaders need to hear from concerned citizens. Sign up for the union of concerned scientists action network to ensure that policy makers get the timely accurate information they need to make informed decisions about global warming solutions ([Ponce, 2009).

Global Warming Control Strategies on International Level

The World Meterological Organisation (WMO) has the responsibility to coordinate the environmental issues related to the atmosphere including air quality, on a global basis. About 7.4, 8.7 and 9.8 per cent of total cumulative CO_2, SO_2 and NO_2 emission respectively could be avoided between 1997 and 2075 by using efficient appliances (Shrestha *et al.,* 1998). Dioxin emissions can be reduced by flue gas treatment by bag filter and flyash treatment by sintering (fusion solidification), chlorination, thermaldehyde and other photolytic and chemical methods (Ahuja, 2008).

Copenhagen Conference

The intention of conference of parties (cop) in Copenhagen was to complete negotiations on a new international agreement on climate change to come into force before 2012. The three page Copenhagen accord managed to secure the unanimous approval of the conference of parties (cop). Point-1 of the Accord, the parties "recognize" the scientific view that the increase in global temperature should be below 2 degree Celsius. Point-2 "deep cuts in global emission are required according to science and as documented by the IPCC Fourth Assessment Report with a view to reduce global emissions so as to hold the increase in global temperature below 2°C, and to take action to meet this objective consistent with the science and on the basis of equality. The final accord was hammered out by the US, China, India, Brazil and South Africa and presented as "take it or leave it" to the rest of the parties, including the EU.

Conclusion and Future Strategies

It is imperative that the developed countries and the rapidly developing countries formulate strategies to curb greenhouse gas emissions. Countries on the fast track of economic growth should also look at adopting new energy saving technologies. It is imperative to carry out a massive reforestation. The emphasis also is laid on increasing the use of renewable energy sources like solar and wind. Application of fertilizer should be reduced.

REFERENCES

Ahmad, Ahsan., Koudstall, Rob., Werners, Saskia, 2006. 'Key Risks". Considering Adaptation to climate change towards a sustainable development.

Ahuja, I. 2008. Global climate change. *Ecol. Modell.* **87**: 51-57.

Alexey, T.M., Hall, J.W. and Lucht, W. 2007. Tipping elements in earth's climate system. *Proceedings of the National Academy of Sciences* **105**: 1786-1793.

Anup, K., Prasad, V.K. and Pant, G.B. 2006. Long-term climate variability and change over monsoon. *Journal of the Indian Geophysical Union* **7**(3): 125-134.

Bindoff, N.L., Willebrand, J., Artale, V. 2007. Oceanic change and sea level. *The Physical Science* **2**: 105-107.

Cannor, S. 2008. Hundreds of methane plumers discovered. *The Independent.*

Choie, O., Brander, K.M. and Erda, L. 2007. Studies show climate change melting permafrost under runways in western arctic weber, Bob Airportbusiness.com.

Christan, P., Wilson, E. and Schneeberger, K. 2003. Modelling changes in the mass balance of glaciers of the northern hemisphere. *Journal of Hydrology* **282**: 145-163.

Crutzen, P.J. 1993. International Society for Optimal Energy **1715**: 230-248.

Energy Information Administration. The Guardian.

Global Warming: Impacts: Forests. United States Environmental Protection Agency, 2007.

Global Warming and its Impact. Wikinews. September 2007, http: /en.wikinews. org.in.

Hitz, S., Smiths, J. and Menne, B. 2007. Estimating global impacts from climate change, p. 66. http: //www.oecd.org.data.oecd.pdf.in.

Jan Sample, 2005. Warming hits 'tipping point'. *Guardian.*

Lau, W.K.M. 2006. Aerosols may cause anomalies in the Indian monsoon. The climate and radiation branch at NASA's Goddard Space Flight Centre. NASA. http: // climate.gsfc.nasa.gov/viewimage.retrieved 2007-04-17.

Lawrence, D.M., Slater, A.G. and Tomas, R.A. 2008. Accelerated artic land warming and permafrost degradation during rapid sca ice loss. *Geophysical Research Letter* **35**: 1506.

Lister, S. 2009. Professor Anthony Costello – Climate Change Biggest threat to humans. *The Times.*

Lobell, D.B., Burke, M.B. and Telbaldi, C. 2007. Prioritizing climate change adaptation needs for food security in 2030. *Science* **319**: 607-610.

Mackenzie, D. 1997. *Forest Fab.* **156**: 15.

Mauri, S. and Pelto, K. 2007. Recent retreat of north-cascade glaciers and changes in north cascade stream flow. *North Cascade Glacier Climate Project.*

Mishra, S., Ramesh, R., Amit, R., Lazar, B., Rajaguru, S.N. and Sandler, A. 2007. High-resolution holocane environmental changes, Northwestern India. *Science* **284**(5411): 125.

Peter, A. Stott, Stone, D.A. and Allen, M.R. 2008. Climate change linked to spread of disease. *Nature* **432**: 610-614.

Ponce, V. Migue 2009. The thirty three facts about global warming. *Nature* **510**: 140-148.

Ramanujan, K. 2004. Forst Glacier in Greenland Doubles Speed. NASA.

Rao, A. and Sinha, A.K. 1994. Climate changes and agriculture. *Nature* **407**: 102-109.

Roy, D'Silva. 2007. The effect and consequences of global warming **5**(2): 11-15.

Royal Society, 2005. Impact of climate change on crops worse than previously though.

Sabine, T., Sarah, G. and Christopher, L. 2004. The ocean sink for anthropogenic CO_2. *Science* **385**(5682): 367-371.

Saseendran, R.M., Smith, I.M.a nd Matson, P.A. 2000. Ecological and evolutionary responses to climate change. *Science* **284**: 1943-1947.

Shah, N., Olsen, S.M and Prasad, K. 2005. Long-term ocean oxygen depletion in response to CO_2 emissions from fossil fuels. *Nature Geoscience* **2**: 105-109.

Shrestha, R.M., Natarajan, B., Chakravarti, K.K. and Shrestha, R. 1998. *Energy Oxford* **23**: 1065-1072.

Sinha, A.K. and Swaminathan, M.S. 1991. Long-term climate variability and changes. *Journal of Indian Geographical Union* **7**(3): 125-134.

Sinha, A.K. and Swaminathan, M.S. 1991. Long-term climate variability and changes. *Journal of Indian Geographical Union* **7**(3): 125-134.

Smith, J., Hitz, S., Akhter, R. 2002. Climate change and Human Health-Risk and Responses. WHO Greeneua.

Stefan, R., Michal, M. and Rasmus, B. 2007. Hurricans and global warming. *Science* **2**: 95-98.

Thukral, A.K. and Virk, G.S. 2001. Environment Protection. *Indian Journal of Environmental Prot.* **13**: 358-367.

Vladimir and Romanovsky, 2007. Climate changes and retreat of glaciers. *Naure* **377**: 687-688.

Zarin, A.S. 2007. Global climate changes and its effect on agriculture. *Nature* **408**: 184-188.

Ziska, E., Fraser, D., Falcon, P. 1997. Assessing risks of climate variability and climate change for rice. *Science* **240**: 996-1002.

Chapter 2

Plant Genetic Resources and Strategies for their Conservation

Salib Shafi, Shabir Ah. Wani, Sanna Parvez and Auqib Hussain Ganie

College of Temperate Sericulture Mirgund, Sher-e-Kashmir University of Agricultural Sciences and Technology of Kashmir, J&K, India

Plant genetic resources are plant genetic materials of actual or potential value. According to the food and agriculture organisation(FOA), plant genetic resources are defined as the entire generative and vegetative reproductive material of species with economical value and /or social value, especially for the agriculture of the present and the future, with special emphasis on nutritional plants. Plant genetic resource conservation has become increasingly important as more plants have become threatened or rare. At the same time, an exploding world population and rapid climate change have led humans to seek new resilient and nutritious crops. Plant genetic conservation strategies generally combine elements of conservation on farm, in-situ, ex situ.

A key event in the conservation of plant genetic resources was the establishment of the international Board for Plant Genetic Resources (IBPGR) (Now Biodiversity International) in 1974, whose mandate was to promote and assist in the world wide effort to collect and conserve the plant germplasm needed for the future research and production. Over recent decades, agriculture has undergone enormous changes, on the one hand, yields per unit area have increased dramatically through a combination of improved crop varieties and a greater use of external inputs. On the other hand, there has been increasing pressure on land for uses other than the production of food, as well as growing concerns about the sustainability and safety of some modern practices.

Genetic resource is a gold mine and we have to make substantial long term investment for exploiting its full potential Without an intelligent and judicious use of PGR it will be difficult to achieve sustainable advances in agricultural production in vitro culture and cryopreservation – important tools to collect and conserve genetic resources in difficult crops. It is time to rethink the approaches, methodologies and focus of PGR to address changing global challenges and to raise the livelihood of resource-poor farmers.

***Keywords**: Conservation, Genetic, PGRs.*

Introduction

Plant genetic resources are plant genetic materials of actual or potential value. They describe the variability within plants that comes from human and natural selection over millennia. Their intrinsic value mainly concerns agricultural crops.

According to the 1983 revised international undertaking on plant genetic resources for food and agriculture of food and agriculture organization (FOA), plant genetic resources are defined as the entire generative and vegetative reproductive material of species with economical value and/or social value, especially for the agriculture of the present and the future, with special emphasis on nutritional plants. Plant genetic resources (PGR) are the most important components of agro-biodiversity. The PGR include primitive forms of cultivated plant species and landraces, modern cultivars, obsolete cultivars, breeding lines and genetic stocks, weedy types and related wild species (IPGRI 1993).

Plant genetic resource conservation has become increasingly important as more plants have become threatened or rare. At the same time, an exploding world population and rapid climate change have led humans to seek new resilient and nutritious crops. Plant genetic conservation stratigies generally combine elements of conservation on farm, *in situ*, *ex situ*. Most *in situ* conservation concerns crop wild relatives, an important source of genetic variation to crop breeding programs. Plant genetic resources that are conserved by any of these methods are often reffered to as germplasm, which is a shorthand term meaning "any genetic materials".

After the second world war efforts to conserve plant genetic resources came mainly from breeders' organisations in the USA and Europe, which led to crop-specific collections primarily located in developed countries. A key event in the conservation of plant genetic resources was the establishment of the International Board for Plant Genetic Resources (IBPGR) (Now Biodiversity International) in 1974, whose mandate was to promote and assist in the world wide effort to collect and conserve the plant germplasm needed for the future research and production. In 2002, the Global Crop Diversity Trust was established by Biodiversity International on behalf of the CGIAR and the FAO through a Crop Diversity Endowment Fund. The goal of the trust is to provide a secure and sustainable source of funding for the worlds most important crop collection.

History of Plant Genetic Resources

The first use of plant genetic resources dates to more than 10000 years ago, when farmers selected from the genetic variation they found in Wild plants to develop their crops. As human population moved to different climates and ecosystems, taking the crops within them, the crops adapted to the new environments, developing for example genetic traits providing tolerance to conditions such as drought, water logging, frost and extreme heat. These traits and the plasticity inherent in having wide genetic variability- are important properties of plant genetic resources.

In recent centuries, although humans had been prolific in collection exotic flora from all corners of the globe to fill their gardens, it wasn't until the early of 20th century. That the widespread and organized collection of plant genetic resources for agricultural use began in earnest. Russian geneticist Nikolai Vavilov, considered by some as the father of plant genetic resources, realized the value of genetic variability for breeding and collected thousands of seeds during his extensive travels to establish one of the first gene banks. He inspired the American Jack Harlan to collect seeds from across the globe for the United States Department of Agriculture. David Fairchild, another botanist at USDA, successfully introduced many important crop deg cherries, soyabeans, pistachios into the United States.

It wasn't until 1967 that the term genetic resources was coined by Otto Frankel and Erna Bennett at the historic International Conference on Crop Plant Exploration And Conservation, organized by the FAO and the International Biological Progra. (IBP). The effective utilization of genetic resources requires that they are adequetly classified and evaluated was a key message from the conference.

Germplasm

The sum total of hereditary material *i.e.* all the alleles of various genes, present in a crop species and its wild relatives. It is also termed as genetic resource. Genetic resource can be broadly grouped into two groups (i) cultivated germplasm (2) wild germplasm depending on the state of of their domestication. Alternatively they may be termed as (1) indigenous (from the country in question) or exotic (from another country) based on their origin. The germplasm of a crop consists of the following five types of materials: (1) land races, (2) obsolete varieties (3) varieties in cultivation (4) breeding lines (5) wild forms and relatives.

Gene Pool Concept

The concept of gene pool was proposed by Harlan and de Wet in 1971. Gene pool consists of all the genes and their alleles present in all such individuals, which hybridize or can be hybridize with each other. In some sense, gene pool describes a concept similar to germplasm. The gene pool of a crop is classified into three groups: (1) primary GP1 (2) secondary GP2 and (3) tertiary GP3 gene pools.

Primary GP1 Gene Pool

GP1 includes all the strains of the concerned crop species. Crossing between members of GP1 is easy, the resulting hybrids are vigorous, show normal meiotic chromosome pairing and recombination, there is normal gene segregation, and seed fertility is complete. Members of primary gene pool are the most commonly used in breeding programmes, and most breeders rely primarily or entirely on this material.

Secondary GP2 Gene Pool

Members of the secondary gene pool are all those species that hybridize with those of the primary gene pool with some to considerable difficulty and the hybrids are at least partially fertile. The difficulty in hybridization is due to ploidy differences, chromosomes alterations or genetic barriers. Gene transfers from GP2 to GP1 are possible, but usually difficult. Members of this group are often used in breeding programmes.

Important Active Germplasm Collections Maintained in India at institutions other than NBPGR, New Delhi

Institution	*Germplasm Collection*
Central Institute for Cotton Research, Nagpur	8768 entries of cottons (2005)
Central Plantation Crops Research Institute, Kasaragod	Plantation crops; 522 accessions
Central Potato Research Institute, Simla	Potato; 2500 accessions
Central Tobacco Research Institute, Rajahmundry	Tobacco; 1500 accessions
Central Tuber Crops Research Institute, Thiruvananthapuram	Tuber crops other than potato; 3871 accessions
CRRI, Cuttack	More than 30000 entries of rice
Directorate of Oilseeds Research, Hyderabad	Oilseeds; 10550 accessions
Directorate of Wheat Research, Karnal	Wheat and barley; 7000 accessions
Project Directorate on Maize, I.A.R.I., New Delhi	Maize; 2500 accessions
Indian Grassland and Fodder Research Institute, Jhansi	Forage and fodder crops; 6267 accessions
Indian Research of Horticultural Research, Bangalore	Tropical fruits; 11467 accessions
Indian Institute of Pulses Research, Kalyanpur, Kanpur	Pulses; 5021 accessions
National Research Institute for Groundnut, Junagarh	Groundnut; 6432 accessions
National Research Centre for Sorghum, Hyderabad	Sorghum; 7366 accessions
National Research Centre for Soybean, Indore	Soybean; 2500 accessions
Sugarcane Breeding Institute, Coimbatore	Over 5861 entries
Indian Institute of Vegetable Research, Varanasi	Vegetables; 16139 Tropical fruits; 11467 accessions

Tertiary GP3 Gene Pool

The species belonging to this group represent the extreme outer limit of the potential germplasm. They cross with the members of primary gene pool with considerable to great difficulty, and hybrids, if produced, are anomalous, lethal or completely sterile. Gene transfers from GP3 to GP1 are extremely difficult and require special techniques, while those from GP3 to GP2 are relatively easier. GP3 is used only occasionally in breeding programme, and that too by a group of researchers having the competence and the patience for tackling the associated problems.

Importance of Plant Genetic Resources

- ☆ Plant genetic resources are playing an ever-growing role on world food security and economic development.
- ☆ They underpin the ability of agriculture to cope with changes, weather environmental or socio-economic.
- ☆ Basic raw material for genetic improvement including designer species
- ☆ Reservoir of useful genes
- ☆ Critical component for food, nutrition, environmental and household security

Facts about India Regarding Plant Genetic Resources

- ☆ One of the 12 world mega biodiversity centers and 17 mega diverse nations
- ☆ One of the 8 centers of origin of crop plants
- ☆ Three of the 34 Hot Spots of Biodiversity- (Himalayas, Indo-Burma, Western Ghats)

Role of Plant Genetic Resources in Increasing Food Production

Over recent decades, agriculture has undergone enormous changes as a result of both technological advances and changing human needs and desires. On the one hand, yields per unit area have increased dramatically through a combination of improved crop varieties and a greater use of external inputs. On the other hand, there has been increasing pressure on land for uses other than the production of food, as well as growing concerns about the sustainability and safety of some modern practices. In spite of advances in food production, food insecurity and malnutrition are still widespread. The latest FAO figures indicate that in 2009 there were around 1 billion chronically hungry people in the world, an increase of about 200 million since the World Food Summit in 1996. It is estimated that the number of hungry people increased by over 100 million due to the food price crisis of 2007-2008 alone. Most of the worst affected people (about 75 percent) live in rural areas of developing countries and depend directly or indirectly on agriculture for a large part of their livelihoods. A 70 percent increase in world agricultural production over today's levels will be required to meet the food demands of the estimated 9.2 billion people in 2050. A major share of this productivity increase will have to come from the use of PGRFA to produce higher yielding, more nutritious, more stable and more eco-efficient crop varieties.

In 2000, the United Nations Millennium Declaration was adopted, committing nations to a new global partnership to reduce extreme poverty and setting out a series of time-bound targets, with a deadline of 2015, that have become known as the Millennium Development Goals (MDG). All countries and all of the world's leading development institutions have agreed to these goals, two of which, in particular, will require the conservation and use of PGRFA if they are to be reached: the eradication of poverty and hunger and the achievement of environmental sustainability. The aim of this chapter is to discuss the role and contribution of PGRFA to food security, sustainable agriculture, economic development and poverty alleviation. The chapter will not review or interpret these four concepts or their inherent complexity and interlinkages. Instead, it will look at the role of PGRFA in the context of some of the emerging and difficult challenges now facing agriculture. Unlike the other seven chapters, this one does not have a counterpart in the first SoW report and so there is no baseline upon which to build. It thus aims to provide an overall review of the current status of PGRFA in relation to sustainable agriculture, food security and economic development, concludes with a summary of some of the main changes that have occurred in recent years and identifies some of the key gaps and needs for the future.

PGRFA plays an important role in all of the four categories. In addition to being a direct 'provisioning service', genetic resources provide the raw material for improving the production of more and better food, either directly or through providing better feed for livestock. They are also important as the basis for improving fibre, fuel or any other crop product. In the area of 'regulating services,' PGRFA are the basis for improving such services as carbon sequestration by crops, for example, deeper-rooted rangeland species and the control of water run-off and soil erosion. The diversity of traditional crops and foods can provide an important cultural service, *e.g.* through its importance in agrotourism or ecotourism; and as a 'supporting service' PGRFA can underpin the development of new varieties, for example food and forage legumes, having an enhanced ability to recycle nutrients such as nitrogen within an agro-ecosystem. In recent years, many programmes have been initiated that seek to enhance these services, in particular, through rewarding those responsible for managing the underlying resource through PES schemes. However, implementing PES is a challenge as many of the services arise from complex processes, making it difficult to determine which actions affect their provision, who is responsible for these actions and who are the beneficiaries who should pay for them. This is particularly true in the case of agro biodiversity. If, for instance, the on farm conservation of a particular traditional crop variety is considered eligible for PES, the challenge is to determine which farmer or farmers should be compensated for its conservation. How much should they receive, for how long, who should pay and what mechanisms are in place for monitoring and ensuring that payments are actually made and that the expected service is actually provided? This is a dilemma that also underlies the debate over how to implement farmers' rights. Nevertheless, PES raises hopes and expectations for the development of a more environmentally-friendly agriculture and the PGRFA sector has a critical role and a responsibility to be part of the debate and action.

Various Activities Involved in PGR Management

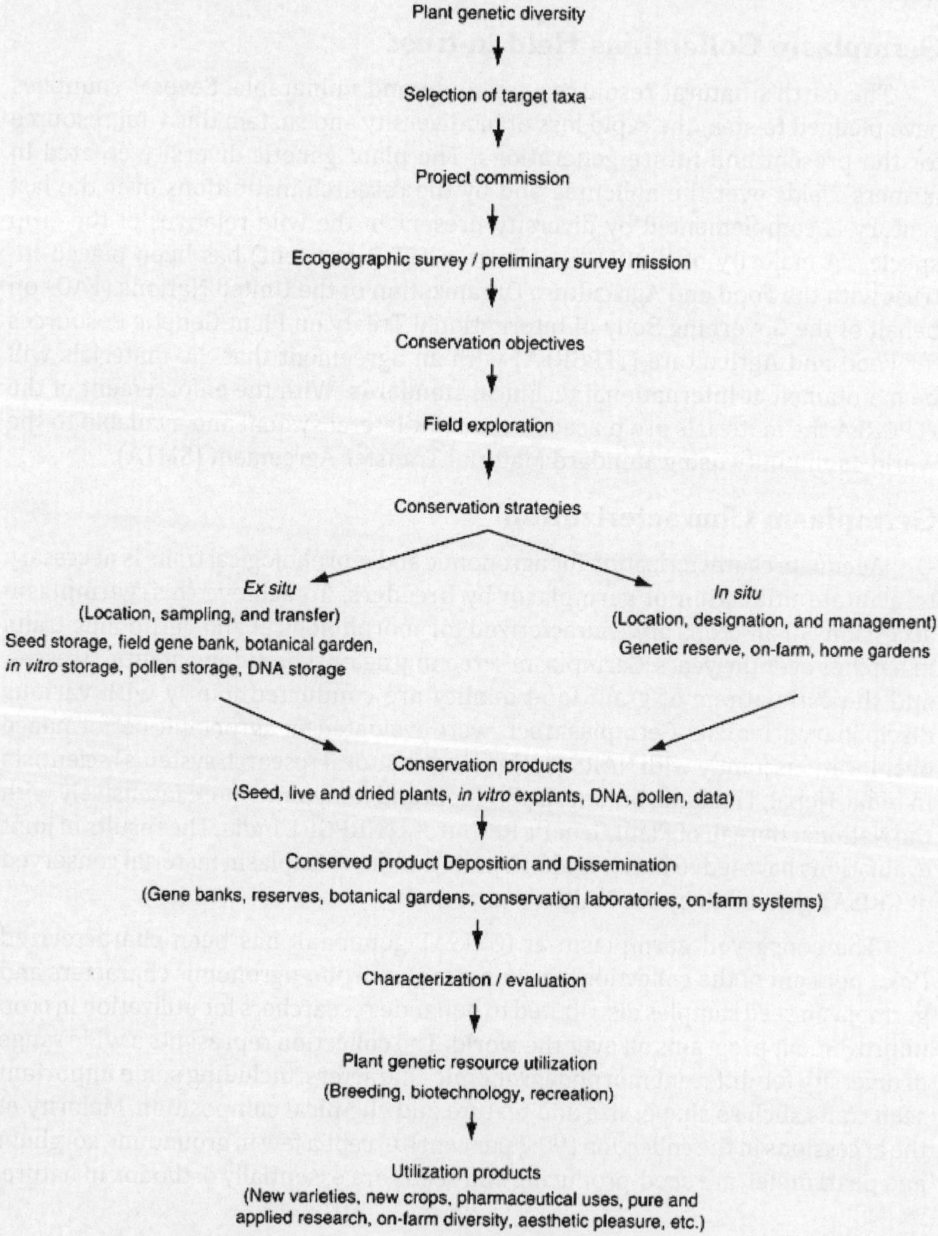

Collection

The collection includes landraces, non-domesticated species, advanced and old cultivars and breeding lines. Information on coordinates of sites of collection is documented on accessions. The collection represents both insurance against genetic erosion and as sources of resistance/tolerance to diseases and pests, climatic and other environmental stresses, improved quality and yield traits for crop improvement.

Germplasm Collections Held in-trust

The earth's natural resources are finite and vulnerable. Several countries have pledged to stem the rapid loss of biodiversity and sustain this vital resource for the present and future generations. The plant genetic diversity created in farmers' fields over the millennia and by the research institutions over the last century is complemented by diversity present in the wild relatives of the crop species. A majority of ICRISAT's collection (95.8 per cent) has been placed in-trust with the Food and Agriculture Organization of the United Nations (FAO) on behalf of the Governing Body of International Treaty on Plant Genetic Resources for Food and Agriculture (ITPGRFA) with an agreement that the materials will be maintained at international technical standards. With the enforcement of the ITPGRFA the materials are placed in the multilateral system and available to the world community using Standard Material Transfer Agreement (SMTA).

Germplasm Characterization

Adequate characterization for agronomic and morphological traits is necessary to acilitate utilization of germplasm by breeders. To achieve this, germplasm accessions of all crops are characterized for morphological and agronomic traits in batches over the years. Germplasm screening against biotic and abiotic stresses, and the estimations of grain food quality are conducted jointly with various disciplinary scientists. Germplasm sets were evaluated for agronomic performance over locations jointly with NARS (national agricultural research systems) scientists in India, Nepal, Thailand, Indonesia, Ethiopia and Kenya and more intensively with the National Bureau of Plant Genetic Resources (NBPGR), India. The results of joint evaluations have led to better understanding of the germplasm material conserved at CRISAT genebank by the NARS scientists.

The conserved germplasm at ICRISAT genebank has been characterized (98.5 per cent of the collection) for important morpho-agronomic characters and germplasm seed samples distributed to bonafide researchers for utilization in crop improvement programs all over the world. The collection represents a wide range of diversity for different morphoagronomic characters, including some important seed traits such as shape, size and texture and chemical composition. Majority of the accessions in the collection (99.7 per cent), except a few in groundnut, sorghum and pearl millet, are seed-producing and seeds are essentially orthodox in nature.

Germplasm characterization is the recording of distinctly identifiable characteristics, which are heritable. This needs to be distinguished from preliminary evaluation, which is the recording of a limited number of agronomic traits considered to be important in crop improvement. Germplasm characterization is carried out in precision fields by spaced planting under adequate agronomic conditions and plant protection. For each accession several morpho-agronomic traits are recorded using the descriptors developed in collaboration with Bioversity International [formerly IPGRI (International Plant Genetic Resources Institute)]. Following these procedures, majority (96.3 per cent) of the germplasm collection at ICRISAT genebank has been characterized. Systematic description of each accession will eventually lead to classification in small and well-organized sectors that will facilitate enhanced utilization of germplasm. The major objectives of germplasm characterization are:

- ✰ Describe accessions, establish their diagnostic characteristics and identify duplicates;
- ✰ Classify groups of accessions using sound criteria;
- ✰ Identify accessions with desired agronomic traits and select entries for more precise evaluation;
- ✰ Develop interrelationships between, or among traits and between geographic groups of cultivars;
- ✰ Estimate the extent of variation in the collection. Through intensive field and laboratory screening and purification, a wide range of sources for desirable traits were identified in the assembled germplasm.

Documentation

Documentation is essential for a genebank management to allow efficient and effective use of germplasm. Characterization and evaluation data are of little use if they are not adequately documented and incorporated into an information system that can facilitate access to data. Information plays a significant role in biodiversity conservation. Accurate information about conserved materials is essential for greater use. Computerized documentation systems enable rapid dissemination of information to users as well as assist curators to manage the collections more efficiently. Tools like GIS and satellite imagery help in searching for germplasm with specific characteristics, monitoring changes in crops and varieties, or deciding where to locate an *in situ* reserve. The Genebank Information Management System (GIMS) of ICRISAT is a user-friendly module designed to integrate various documentation activities and provides information on accessions due for regeneration/viability monitoring at any given point of time.

Germplasm Utilization

Germplasm conserved at ICRISAT genebank has become an important source of diversity available to researchers in both public and private sectors throughout the world. For example, between 1975 and 2007, ICRISAT genebank has distributed over 686,000 samples of its mandate crops and small millets to users in 144 countries. The global collections held at ICRISAT serve the purpose of restoration of germplasm to the source countries when national collections are lost due to natural calamities, civil strife, *etc.* We supplied 362 sorghum accessions to Botswana; 1827 sorghum and 922 pearl millet to Cameroon; 1723 sorghum and 931 chickpea to Ethiopia; 838 sorghum and 332 pigeonpea to Kenya; 1436 and 445 sorghum accessions to Nigeria and Somalia, respectively; 71 pigeonpea accessions to Sri Lanka and 44,822 accessions of ICRISAT mandate crops to NBPGR, India. Thus the national programs of several countries have regained their precious plant germplasm heritage which could have been lost if this was not onserved in the ICRISAT genebank.

Core Collections

As defined by Brown, a core collection is a selected and limited set (5–20 per cent of the total collection) of accessions derived from an existing germplasm collection, chosen to represent the genetic spectrum in the whole collection (reserve collection) and including as much as possible of its diversity. Core collections were initially proposed by O. Frankel and A.H.D. Brown in 1984 to address the issue of large germplasm collections (Brown, 1995).

Pre-breeding

- ☆ Pre-breeding is the step before practical breeding.
- ☆ Its aim is "To introduce new desirable traits/genes into an adapted genetic background. This will broaden the genetic base in a breeding material in pace with environmental changes."
- ☆ A vital contribution of pre-breeding is increasing the total genetic diversity in crops and finding specific genes and traits.

It is vital link between conservation of PGR in gene bank collections and utilisation of these resources in agriculture and horticulture.

Application of Biotech Tools

- ☆ Exploration and Collection: Tissue culture, Molecular markers
- ☆ Conservation: Molecular markers,

  Tissue culture and cryo-preservation
- ☆ Genomic resources: Gene cloning, genetic engineering, molecular markers
- ☆ Quarantine: Molecular markers for pathogen detection
- ☆ Utilization of germplasm- marker techniques, embryo rescue (prebreeding), cloning

PGR Management through Molecular Markers

- ☆ Isozyme markers
- ☆ RFLP
- ☆ RAPD
- ☆ ISSR
- ☆ AFLP
- ☆ Microsatellite markers
- ☆ Single Nucleotide Polymorphism (SNPs)
- ☆ Sequence based markers

Use of Molecular Markers in PGR Management

Diversity Analysis of Germplasm

- ☆ Phylogenetic relationship
- ☆ Development of core collection
- ☆ Gene flow study

DNA Fingerprinting

- ☆ Germplasm identification
- ☆ Genetic purity/Genetic stability
- ☆ Identification of duplicates

Gene Discovery

- ☆ Association mapping
- ☆ Allele mining

Trait Specific Germplasm

- ☆ Identification and Validation

DNA Fingerprinting

DNA fingerprinting protocols have been developed using STMS, AFLP, ISSR, RAPD and SRAP techniques in 34 crops and 2215 cultivars have been fingerprinted.

Specific molecular markers for seed purity testing of commercial hybrids of cotton, pearl millet and sorghum and identification of citrus rootstocks developed

National Genomic Resource Repository

- ☆ It comprises of genomic DNA including DNA of rare and endangered species, genomic and cDNA libraries, BAC libraries, gene constructs and promoters *etc.*
- ☆ A single-window system for sharing genomic resources

- ☆ Efficient utilization of genomic resources
- ☆ A mechanism to check biopiracy

Conclusion

Genetic resource is a gold mine and we have to make substantial long term investment for exploiting its full potential Without an intelligent and judicious use of PGR it will be difficult to achieve sustainable advances in agricultural production *in vitro* culture and cryopreservation - important tools to collect and conserve genetic resources in difficult crops Molecular markers techniques - strong tool for PGR management in conservation, collection, utilization of PGRs. It is time to rethink the approaches, methodologies and focus of PGR to address changing global challenges and to raise the livelihood of resource-poor farmers.

REFERENCES

Ananda Rao, A., Chauhan, S.S., Radhakrishnan, R., Tikader, A., Borpuzari, M.M., Kamble, C.K., 2011. Distribution, variation and conservation of mulberry (*Morus* spp.) genetic resources in the arid zone of Rajasthan, India. Biorem. Biodiv. Bioavail. 5, 52–62.

Arora, R.K., 1991. Plant diversity in the Indian gene center. In: Paroda, R.S., Arora, R.K. (Eds.), Plant Genetic Resources: Conservation and Management. NBPGR, New Delhi, pp. 25–44.

Chandrasekhar, M., Saraswat, R.P., Kamble, C.K., 2009. Strategies for conservation of mulberry genetic resources (*Morus* spp.) in India. In: Proceedings of National Workshop on Seri-Biodiversity Conservation CSGRC, Hosur, India, pp. 23–27.

Dodds, J.H., Roberts, L.W., 1985. Experiments in Plant Tissue Culture, 2nd ed. Cambridge University Press, London, pp. 172–179.

Chapter 3

Exploration of Locally Available Medicinal Plants and their Efficacy on Silkworm Diseases

Shabir A. Bhat[1], Ifat Bashir[2], F.A. Malik[1], M.Y. Khan[1], Z.I. Buhroo[1], S.A. Wani[1] and Shahnawaz Ahmad[2]

[1]College of Temperate Sericulture SKUAST-Kashmir (Mirgund), J&K
[2]Govt Degree College for Boys Sopore Kashmir, J&K

Various locally available medicinal plants were collected, processed, pulverized and their extracts (aqueous/methanol) were tested against silkworm microsporidiosis and Bm nuclear polyhedrosis both in vitro and in vivo. Perusal of results revealed extracs irrespective of plant were effective in lower down the mortality due to microsporidiosis and nuclear polyhedrosis. However, Ammi copticum, Citrus paradesi at all concentrations, Phyllanthus niruri at 2.00 and 3.00, Myrtus caryophyllus, Lawsonia alba at 3.00 per cent controlled microsporodiosis incidenceto the tune of 100 per cent, however, Viscum album at 1.50 per cent performed comparatively better in lowering down the larval and pupal mortality in nuclear polyhedrosis.

***Keywords**: Microsporidians, Concentration, Nuclear polyhedrosis, Plant etc.*

Introduction

Mulberry silkworm, *Bombyx mori*, L. is reared in colonial form past several centuries under controlled environmental, nutritional and germ free conditions which has resulted in loss of some of the ancestral characters including the resistance against various infectious diseases which are broadly classified into

4-types (Table 3.1). Virosis is sub-divided into nuclear polyhedrosis, cytoplasmic polyhedrosis, infectious flacherie and densonucleosis caused by *Bombyx mori* nuclear polyhedrosis virus, *Bombyx mori* cytoplasmic polyhedrosis virus, *Bombyx mori* infectious flacherie virus and *Bombyx mori* densonucleosis virus respectively. Mycosis such as white, green, yellow and red muscardine caused by *Beauveria bassiana, Metarhizium anisopliae, Pacilomyces farinosus* and *Sporosporella urella* respectively. The aspergillosis also comes under fungal disease and occurs during chawki rearing only and is caused by *Aspergillus flavus, A. tamarii*. The bacterimia such as bacterial disease of digestive tract, septicemia and bacterial toxicosis

Table 3.1: Infectious Diseases of the Silkworm *Bombyx mori*, L.

Diseases	*Pathogencity*	*Symptoms*
	Viruses	
Bombyx mori Nuclear polyhedrosis	Chronic	Swell on inter segmental region shinning and fragile skin, milky white fluid.
Bombyx mori Cytoplasmic polyhedrosis	Chronic	Translucent cephalothoraxic region; diarrhea; retarded growth; milky white midgut; whitish feaces.
Bombyx mori Infectious flacherie	Chronic	Translucent cephalothoraxes; retarded growth; vomiting and diarrhea.
Bombyx mori Densonucleosis	Chronic	Translucent cephalothoraxes; retarded growth; vomiting and diarrhoea.
	Mycosis	
White muscardine	Chronic	Oily specks on the body surface; larva on death softens, turns hard and latter mummifies; mummified larvae appear white.
Green muscardine	Chronic	Large specks with black periphery; mummified larvae green in color
Yellow muscardine	Acute/ Chronic	Large diseases specks around stigma and small on skin, mummified larvae yellow.
Red muscardine	Acute/ Chronic	Develop red patches few hours before death; no external growth.
	Bacterimia	
Bacterial diseases of digestive organ	Chronic	Sluggish movement; retarded growth; transparent cephalothoracic region
Septicemia	Acute	Sluggish movement low appetite, swollen thorax, shrinkage; vomiting softening and discolored body.
Toxicosis	Acute/ Chronic	Sluggish movement; retarded growth; cessation of feeding; vomiting; paralysis and death; corpse stretch and cephalothoracic region bend like hook.
Microsporidiosis	Chronic	Flaccid larvae; retarded growth; white pustules all along the length of the silk gland

caused by *Bacillus* sp., *Streptococci* sp., *Stephylococci* sp., *Serratia marcescens* and *Bacillus thringiensis* respectively. The microsporidiosis commonly called as pebrine caused by *Nosema bombycis* and other strains is most dreaded disease in silkworm. The disease is transovarially transmitted and has vanished silk industry in most western countries in past and has also eliminated famous Kashmiri univoltine race. The annual cocoon crop loss due to these diseases was estimated to be over 30-40 per cent (Nataraju *et al.*, 2005). Illahi and Nataraju (2007) reported 13.8-26.0 per cent loss due to nuclear polyhedrosis resulting in yield reduction from 11.0 - 17.8kg per 100 disease free layings. The incidence due to pebrine disease ranged between 18.98, 13.46 and 9.00 percent in south, north and central zone of Kashmir respectively (Ganie *et al.*, 2008). Normally microsporidiosis occurrence is of low intensity during summer and high in winter season in tropics due to the increase in the lepidopteron insects during winter and rainy seasons. However due to the advancement in silkworm crop protection technologies the disease level has come down to 15-20 per cent. Various prophylactic measures including use of botanicals like *Psoralea coryleifolia* and *Tribulus terrestris* (Sameul Manoharraj, 1994), *Acacia suma, Caesalpinia coriaria* and *Terminalia tomentosa* (Manoharan, 1996) were reported to reduce the incidence of grasserie disease when the aqueous extracts of these plant products were applied on mulberry leaves before feeding to the silkworms. *Ammi copticum and Viscum album* have been used as a traditional medicine and has been also tested in silkworm diseases (Bhat and Nataraju, 2006; Khwaja *et al.*, 1986). Many workers (Hostanska *et al.*, 1995 and Frank *et al.*, 2003) have reported that Mistletoe has physiological, anti-cancer, anti-viral and immunostimulatory properties.

Material and Methods

To develop eco and user friendly approach in management of silkworm microspodiosis and nuclear polyhedrosis various locally available plants/plant products such as seeds of *Ammi copticum* (Ajowan) of family *Umbelliferae, Citrus paradesi* (Grape seed) of family *Rutaceae,* Pedicle of *Myrtus caryophyllus* (Clove) of family *Myrataceae*; Leaves of *Phyllanthus niruri* (Phyllanthus) of family *Euphorbiaceae;* leaves of *Lawsonia alba* (Mehandi) of family *Lythraceae,* rhizome of *Curcuma longa* (Turmeric) of family *Zingiberaceae* and leaves and stem of Misteltoe *Viscum album* (Aal) of family *Santalaceae* (Plate 1.1) were collected and used in present study. The plant materials were washed thoroughly with distilled water, dried in shade and powdered by using mixer and grinder. The powder was then dissolved in either distilled water/methanol and the solution was filtered using absorbent cotton/Whatman's filter paper. The final extract was then dissolved in distilled water and diluted to 1.00, 2.00 and 3.00 per cent for *A. copticum, C. paradesi, M. caryophyllus, P. niruri, L. alba* and *C. longa* respectively and 1.00, 1.50 and 2.00 per cent for *V. album* and stored at 5°C till use.

The required concentration (1×10^6 microsporidian spores/POB/ml)of microsporidian spores/*Bombyx mori* NPV were prepared from the stock solution

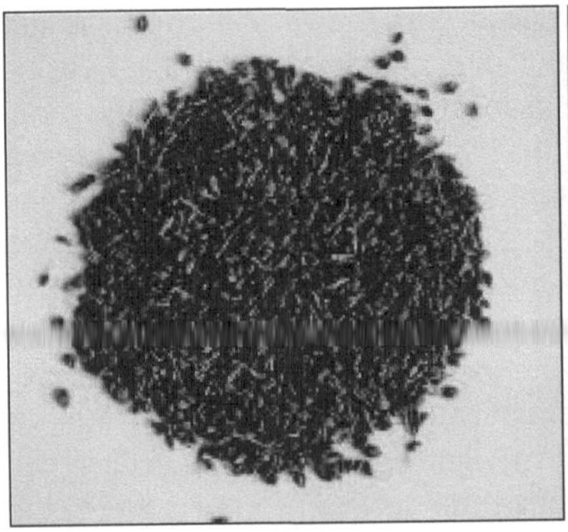

Ajowan Seeds

Grape Seeds

Clove Pedicle

Phyllanthus Plant

Mehandi Leaves

Rhizome Turmeric

Plate 3.1: Different Medicinal Plants Utilized in the Study.

Misteltoe on Walnut Plant as Parasite

Misteltoe Leaves after Collection

Plate 3.1: Different Medicinal Plants Utilized in the Study.

and quantified by following the standard Neubauer's haemocytometer method (Cantwell, 1974).

Results and Discussions

The highly purified pathogen of microsporidian spores or nuclear polyhedral bodies of ($1x10^6$) were suspended in one ml of botanical separately and kept for 30 minutes at room temperature and after 30minutes of incubation, the suspensions were centrifuged and the sediment containing the incubated microsporidian spores/polyhedral bodies were washed in distilled water by repeated centrifugation and observed Microscope for effectiveness, (+) was marked for effective and (-) was marked for ineffective and the results are presented in Table 3.2. In 2nd set of experiment out of 3rd instar silkworms were inoculated with 1ml of known concentration of the pathogens (microsporidian spores or nuclear polyhedral bodies) through *per os*. The mulberry leaf fortified/treated with botanicals were given to the inoculated larvae on alternate days till onset of spinning and the larval mortality if any were recorded acoordingly. Perusal of data revealed aqueous extracts of *A. copticum* and *C. paradesi* at 1.00-3.00 per cent, *P. niruri* at 2.00-3.00 per cent and *M. caryophyllus* at 3.00 per cent were found effective in controlling the mortality to the tune of 100 per cent in microsporidian infected batch, however *V. album* at 1.50 per cent was found most promising among all tested concentration in minimizing the larval mortality as compared to other treatements and inoculated control (Table 3.3). Similarly pupal mortality was also recorded comparatively low in 1.50 concentration in comparison to inoculated batch where (19.00) pupal mortality was recorded (Table 3.4). Thus, *in vivo* screening of these botanicals confirmed the observation of *in vitro*.

Table 3.2: *In vitro* Studies of Certain Medicinal Plants against Microsporidian Spores and Nuclear Polyhedral Bodies

Common Name	*Scientific Name*	*Plant Part Used*	*Concentrations (Per cent)*	*Effect on Spore/ Polyhedral Bodies*
Ajowan	*Ammi copticum*	Seeds	1.00	+
			2.00	+
			3.00	+
Grape seeds	*Citrus paradesi*	Seeds	1.00	+
			2.00	+
			3.00	+
Clove	*Myrtus caryophyllus*	Pedicel	1.00	–
			2.00	–
			3.00	+
Phyllanthus	*Phyllanthus niruri*	Leaves	1.00	–
			2.00	+
			3.00	+
Mehandi	*Lawsonia alba*	Leaves	1.00	–
			2.00	–
			3.00	–
Turmeric	*Curcuma longa*	Rhizome	1.00	–
			2.00	–
			3.00	–
Misteltoe	*V. album*	Leaves and stem	1.00	+
			1.50	+
			2.00	+
Control	–	–	–	–

+: Effective; –: Ineffective

Table 3. Efficacy of botanical extracts on microsporodiosis.

Treatment	*Mortality in different Treatments*		
	1.00 per cent	*2.00 per cent*	*3.00 per cent*
A. copticum	0.00	0.00	0.00
C. paradesi	0.00	0.00	0.00
M.caryophyllus	1.33	1.00	0.00
P. niruri	1.67	0.00	0.00
L. alba	2.00	1.67	0.00
C.longa	2.00	2.00	0.67
Inoculated Control	12.33		

Table 3.4 Efficacy of *Viscum album* Extract on *Bombyx mori* Nuclear Polyhedrosis

Treatment	*Concentrations (per cent)*	*Mortality Due to the Inoculation*	
		Larval	*Pupal*
V. album	1.00	2.0	10.0
	1.50	0.4	4.2
	2.00	1.4	6.6
Inoculated control	-	5.0	19

It is well documented that silkworm diseases cannot be controlled but can be managed to some extent by adopting prophylactic measures like disinfection of rearing house/appliances, chemotherapy, thermotherapy *etc.* The use of botanicals is considered as one of the recent development in the field of silkworm pathology and disease management and is regarded appropriate, farmer friendly and economical means of minimizing the spread of diseases in silkworm *B. mori* population. It has been reported that biomolecules present in botanicals/plant products have both nutritive, antimicrobial, antiviral, antioxidant properties which could be exploited through their application during silkworm rearing. These botanicals act as growth promoting factors and indirectly help in reducing disease spread in the silkworm population (Latha *et al.*, 2011). *V. album* has been reported to have antiviral properties and also enhances the longevity in silk-moths (Kim *et al.*, 2010). Several botanicals experimented on silkworm diseases and revealed that they have phagostimulant, antimicrobial, growth promoter properties *etc.* (Gangadhar, 2004; Shubha, 2005; Sridevi, 2003; Manimegalai and Chandramohan, 2006, Bhat and Nataraju 2006, Nayeema and Bhat 2017). Grape seeds are laden with biologically active ingredients having antibiotic, anti-fungal, antiviral and anti-protozoan activity (House, 1974). In view of the importance of the medicinal plants and their role in minimizing the disease incidence and increase economic traits, various medicinal plants under report were collected from different locations, processed, pulverized and were tested against silkworm, *B. mori* microsporodiosis and Nuclear polyhedrosis through *in vitro* and *in vivo* studies. The plant products *viz., A. copticum, C. paradesi* at all concentrations, *P. niruri* (2.00 and 3.00 per cent) and *V. album* at 1.50 per cent were most found most effective against microsporidiosis and Nuclear polyhedrosis respectively. These plant products not only minimized the diseases but also enhanced the various economic traits of the silkworms. It is therefore concluded, apart from the general management practices with the use of disinfectants (chemical and physical), botanicals such as seeds of *A. copticum, C. paradise*, leaves of *P. nuriri* and leaves/stem of *V. album* besides supplementation could be used as one of the component in integrated disease management (IDM) of silkworm microsporidiosis and Nuclear polyhedrosis.

REFERENCES

Bhat, S. A and Nataraju, B. 2006. Characterization of microsporidian infecting Lamerin breed of the silkworm, *Bombyx mori* L Ph.D Thesis, University of Mysore, Mysore.

Cantwell, G. E. 1974. Insect diseases, vol. II, Marcel Dekkar Inc. New York pp. 501-547.

Frank, U., Engels, I., Wagner, A., Lacour, M. and Daschner, F. D. 2003. Influence of mistletoe (*Viscum album*) extracts on phagocytosis/burst activity of human phagocytes. *European Journal of Clinical Microbial Infect Disease* **22**(8): 501-503.

Gangadhar, M. 2004. Studies on botanicals on silkworm growth and the inhibition of *BM*KPV of silkworm using serological techniques. M. Sc. (Pl. Biotech.) Thesis, UAS, Bangalore, India pp. 78.

Ganie, N.A., Sahaf, K. A., Munshi. N.A., Zargar, M. A. and Baba, Z. A. 2008. Incidence of protozoan diseases (Pebrine) of silkworm *Bombyx mori* L in Kashmir *SKUAST J. Res.* **10:** 134-138.

Hostanska, K., Hajto, T., Spagnoli, G. C., Fischer, J., Lentzen, H. and Hermann, R. 1995. A plant lectin derived from *Viscum album* induces cytokine gene expression and protein production in cultures of human peripheral blood mononuclear cells, *Nat Immun.* **14**(5-6): 295-304.

House, H. L. 1974. Nutrition, In: In the physiology of Insectae, (M. Rockstein, ed.) Second ed. Vol. 2: 43-84. Academic Press New York.

Illahi, I. and Nataraju, B. 2007. Prevalence of to Nuclear polyhedrosis in mulberry silkworm, *Bombyx mori* L In Jammu and Kashmir, *Indian Journal of Sericulture,* **46:** 43-48.

Khwaja, T. A., Dias, C. B. and Pentecost, S. 1986. Recent studies on the anti cancer activities of mistletoe (*Viscum album*) and its alkaloids, *Oncology* **43**(1): 42-50.

Kim, B. K., Choi, M. J., Park, K. Y. and Cho, E. J. 2010. Protective effects of Korean mistletoe lectin on radical-induced oxidative stress, *Biology Pharmology Bulletin* **33**(7): 1152-1158.

Latha, S., Bhaskar, R. N., Pallavi and Chikkalingaiah. 2011. Effect of botanical treatment to *Bm*NPV polyhedral bodies on cocoon parameters of silkworm *B. mori* (PM×CSR_2). *International Journal of Advanced Biological Research* **1**(1): 45-51.

Manimegalai, S. and Chandramohan, N. 2006. Efficacy of botanicals against grasserie disease of silkworm, *Bomby × mori* L. *Sericologia* **46**(1): 15-23.

Manoharan, S. 1996. Evaluation of certain botanicals for the management of grasserie of *Bomby × mori* L. M. Sc. (Agri.)Thesis, TNAU, Coimbatore, India pp. 78.

Nataraju B., Sathyaprased K., Munjunath, D. and Aswath, K. 2005. A Text Book on Silkworm Crop Protection. CSB Ministry of Textiles, GoI. 86-147.

Nayeema, J and Bhat, S. A. 2017. Efficacy of Mistletoe (*Viscum album*) against Nuclear polyhedrosis virus of Silkworm, *Bombyx mori* L. M. Sc. Thesis, Temperate Sericulture Research Institute, Faculty of Agriculture, Sher-e-Kashmir University of Agricultural Sciences and technology of Kashmir. 44-46.

Samuel Manohar Raj, K. 1994. Effect of certain botanicals on the nuclear polyhedrosis virus disease of *Bomby × mori* L. M. Sc. Thesis, TNAU, Coimbatore, India pp. 62.

Shubha, K. 2005. Efficacy of medicinal plant extracts on stability and spread of *Bm*NPV. M. Sc. (Seri.) Thesis, UAS, Bangalore, India pp. 84.

Sridevi, G. 2003. Effect of mulberry leaves fortified with medicinal botanicals on the performance of mulberry silkworm, *Bombyx mori* L. M. Sc (Seri.) Thesis, University of Agricultural Science, Bangalore pp. 116.

Chapter 4

Biodiversity: A Popular Way of Describing the Diversity of Life on Earth

*M.A. Bhat[1] *, Z.I. Buhroo[2], Suraksha Chanotra[1] and Abdul Aziz[1]*

[1]Department of Sericulture, Poonch Campus, University of Jammu, Jammu – 185 101, J&K
[2]College of Temperate Sericulture-Mirgund, Sher-e-Kashmir University of Agricultural Sciences and Technology of Kashmir, Srinagar – 190 025, J&K
**Corresponding author e-mail: mbhat62@gmail.com*

In the developing world, biodiversity provides the assurance of food, countless raw materials such as fiber for clothing, materials for shelter, fertilizer, fuel and medicines, as well as a source of work energy in the form of animal traction. The rural poor depend upon biological resources for an estimated 90 percent of their needs. In the industrialized world access to diverse biological resources is necessary to support a vast array of industrial products. In the continuing drive to develop efficient and sustainable agriculture for many different conditions, these resources provide raw material for plant and animal breeding as well as the new biotechnologies. In addition, biodiversity maintains the ecological balance necessary for planetary and human survival.

Keywords: *Biodiversity, Diversity, Earth, Life.*

Overview

Biodiversity is a popular way of describing the diversity of life on earth: it includes all life forms and the ecosystems of which they are a part. World Food Day – the anniversary of FAO's founding on 16 October 1945 – celebrates, in particular, that part of biodiversity that nurtures people and contributes to long-term food security for all. Biodiversity forms the foundation for sustainable development. It is the basis for the environmental health of our planet and the source of economic and ecological security for future generations. In the developing world, biodiversity provides the assurance of food, countless raw materials such as fibre for clothing, materials for shelter, fertilizer, fuel and medicines, as well as a source of work energy in the form of animal traction. The rural poor depend upon biological resources for an estimated 90 percent of their needs. In the industrialized world access to diverse biological resources is necessary to support a vast array of industrial products. In the continuing drive to develop efficient and sustainable agriculture for many different conditions, these resources provide raw material for plant and animal breeding as well as the new biotechnologies. In addition, biodiversity maintains the ecological balance necessary for planetary and human survival ((Mora *et al.*, 2011).

We are losing biological diversity at an unprecedented rate. The loss of species is not new. In the course of geological time one has only to recall the fate of the dinosaurs. By and large, however, the disappearance of species in past eras has occurred by virtue of natural processes within the context of evolutionary time-scales. Today, however, human activities contribute more to the loss of biodiversity than any other factor. Biological resources are renewable resources, but they are being exploited at rates that exceed their sustainable yield. Human destruction of habitats, whether exploited for commercial or subsistence reasons, is the greatest threat. The clearing of land for agriculture, overgrazing of grasslands, cutting and burning of forests, unsustainable logging and fuelwood collection, indiscriminate use of fertilizers and pesticides, over-watering of crops, overexploitation of fisheries, draining and filling of wetlands, poor water management, urbanization and pollution of air and water, figure prominently in the degradation of our biological resources. Five to seven million hectares of cultivated land are destroyed every year (Field *et al.*, 2009).

Loss of biodiversity is frequently presented as an environmental problem, but the underlying causes are essentially social, economic and political. The excessive and unsustainable consumption of resources by a small but rich minority of the world's population, combined with the destructive impacts of the world's poor and hungry in a desperate bid for survival, have destroyed or overexploited habitats worldwide. Genetic erosion - the reduction of diversity within and the main cause of extinction of a species - is a global threat to agriculture. The greatest loss of crop genetic resources results from the introduction of modern, uniform plant varieties in place of a mix of traditional ones. The Green Revolution introduced high-yielding varieties of rice and wheat to the developing world, but displaced traditional varieties and their wild relatives on a massive scale. In India, for example, agronomists predict that just ten rice varieties will soon cover three-quarters of the total rice area where once over 30 000 different varieties were grown. In the United States, over 85 percent of the 7 000 or so apple varieties grown in the last century are now extinct. The same is true of animal genetic resources. The introduction of a very few modern breeds that are better suited for the high input-output of industrial agriculture is displacing the diversity of indigenous livestock breeds. In Europe, half of all the breeds of domestic animals (horses, cattle, sheep, goats, pigs and poultry) that existed at the beginning of the century have become extinct. A third of the remaining 770 breeds are in danger of disappearing within the next 20 years (Ayoade *et al.*, 2009; Agarwal *et al.*, 2011).

What is Biodiversity?

Biological diversity is made up of all species of plants and animals, their genetic material and the ecosystems of which they are a part. Genetic diversity refers to the variation of genes and genotypes between and within species. It is the sum total of varied genetic information contained in the genes of individual plants, animals and microorganisms that inhabit the earth. Diversity within a species gives the ability to adapt to change whether in environment, climate and agricultural methods or the presence of new pests and diseases (Robosky, 2009).

Species Diversity Refers to the Variety of Species within a given Area

Ecosystems consist of interdependent communities of species (complex mixes of diversity between and within species) and their physical environment. The extent of an ecosystem or habitat is imprecise; a single ecosystem may cover thousands of hectares or just a few. They include major natural systems such as grasslands, mangroves, coral reefs, wetlands and tropical forests as well as agricultural ecosystems that, while depending upon human activity for their existence and maintenance, have characteristic assemblages of plants and animals (Gaston, 2000).

Why is Biodiversity Important?

Genetic diversity in agriculture enables crops and animals to adapt to different environments and growing conditions. The ability of a particular variety to withstand drought or inundation, grow in poor or rich soil, resist one of the many insect pests or diseases, give higher protein yields or produce a better-tasting food are traits passed on naturally by its genes. This genetic material constitutes the raw material that plant and animal breeders and biotechnologists use to produce new varieties and breeds. Without this diversity we would lose the ability to adapt to ever-changing needs and conditions. Sustainable agriculture could not then be achieved in many of the world's different food production environments. Diversity among individual plants and animals, species and ecosystems provides the raw material that enables human communities to adapt to change – now and in the future. Deprived of biodiversity, the ability of humankind to meet the challenges resulting, for example, from global warming and ozone depletion would be severely limited. The diversity found within the small number of plant and animal species which form the basis of world agriculture and food production remains a small but vital part of the earth's biodiversity. Through modern biotechnologies wild diversity can also be incorporated into crops and contribute to world agricultural development (IUCN: facts and figures on biodiversity, 2012).

Biodiversity for Sustainable Development

The biological resources of each and every country are important, but not all are equally endowed. In general, a small number of countries lying within the tropics and subtropics account for a very high percentage of the world's biodiversity. Tropical forests, for example, cover only 7 percent of the earth's land surface, but they are estimated to contain at least 50 percent of all species. The most important food crops, however, appear to have originated in areas that have pronounced seasons, not in the tropical forests. This tends to coincide with arid and semi-arid zones, which include famine-prone countries such as Ethiopia. It makes sense, therefore, to look for sources of certain food crop diversity in such areas. A single Ethiopian barley plant, for example, has yielded a gene that now protects California's annual barley crop from yellow dwarf virus. The fact that the richest nations are home to the smallest pockets of biodiversity while the poorest are stewards of the richest reservoirs underscores the interdependency of all nations, and the urgency of crafting common strategies for sustaining biodiversity that share both responsibility and benefits. On the eve of the twenty-first century, the challenge for the global community is not to save biodiversity for its own sake, but to ensure that biodiversity is used sustainably and equitably for human development (Dalal-Clayton and Bass, 2002).

Indigenous Knowledge and Biodiversity

The wealth of living things on the earth is the product of hundreds of millions of years of evolutionary history. Since the emergence of Homo sapiens from the ranks of humanoid primates, biodiversity and humanity have become inextricably linked. Human cultures have adapted too many diverse habitats. They have used, altered and nurtured biological resources to meet countless needs. As a result of plant and animal domestication, and resource harvesting, a tremendous interdependence has evolved between "natural" and "human-induced" biodiversity. For centuries rural peoples have encouraged and relied upon biodiversity for their livelihoods. Farmers have managed genetic resources for as long as they have cultivated crops. For some 12 000 years, they have selected varieties of crops and livestock breeds to meet environmental conditions and diverse nutritional and social needs. The immense genetic diversity of traditional farming systems is the product of human innovation and experimentation – both historic and ongoing. This has been recognized in FAO by the resolution on farmers' rights that acknowledges the past, present and future contributions of farmers in conserving, improving and making available plant genetic resources, and that they should be rewarded for their contributions. For a large number of developing countries, self-reliance in food production will depend on improving low-input agriculture in difficult environmental conditions. The raw materials for these improvements are the biological resources sustained in forests, rangelands, fields and farms. The accumulated knowledge of farmers, coupled with access to modern technologies, provides the key to developing sustainable agricultural systems. In many parts of the world, wild species and natural habitats still help support household food security- access by all people at all times to the food they need for a healthy life. In Nepal, for example, 135 tree species are used as fodder. In Ghana, three-quarters of the population look to wildlife for most of their animal protein. Access to advanced science and technology contributes to reducing human suffering and promoting economic development. But there is a growing recognition of the value of indigenous knowledge to address global agricultural, health and environmental problems. And there is increasing awareness that conservation and use of biodiversity must be concerned not only with genes, genotypes, species and ecosystems, but also with the traditional knowledge that has helped to produce and maintain this diversity (Alabi, 1992).

Traditional Medicines

Estimated three-quarters of prescription drugs derived from plants were discovered because of their prior use in indigenous medicine. Forest-dwelling indigenous people employ at least 1 300 plant species for medicines and related purposes. Over 60 species of plants are used to treat skin infections in the Amazon region alone (Shaw, 1998).

Traditional Food Plants

For generations, subsistence farmers have been producing or gathering plants in the wild or semi-wild that have long been accepted as desirable sources of food. At least 1 000 million people are estimated to use such traditional plants to satisfy their food needs. They are essential to the diets of rural subsistence households throughout the developing world, providing sources of energy, vitamins and minerals. Many countries, however, are experiencing a shift away from traditional foods, resulting in a narrowing of the food base. Failure to appreciate their benefits, together with a growing demand for imported foods, is reducing availability and consumption of these foods. Since 1985, FAO, in cooperation with governments, has been endeavoring to reverse this trend. In association with research institutions, FAO is also promoting genetic improvements in traditional plants and development of technologies for preparing foods based on these plants that are acceptable in urban markets. Most rural communities in Botswana depend primarily on arable crops as their source of food. Globally, studies have shown that rural communities face food insecurity and are chronically malnourished (Tiisekwa *et al.,* 2004) because of drought stress (Boyer, 1982; Ludlow and Muchow, 1990; Harris and Mohammed, 2003; Babu, 2000), low adoption level of improved crop production technologies (Babu, 2000),

Farming System Diversity in the Developing World

Four broad agro-ecological zones account for 90 percent of agricultural production in developing countries: drylands and areas of uncertain rainfall; humid and per-humid lowlands; irrigated areas; and hill and mountain areas. Each of these zones has a range of farming systems and a mixture of traditional and modern intensive production systems (John Dixon and Aidan Gulliver and David Gibbon, 2001).

Biotechnology and Biodiversity

Biotechnology is generally considered to be "any technique that uses living organisms to make or modify a product, to improve plants or animals, or to develop microorganisms for specific uses". Modern biotechnologies offer vast potential for improving the quality and increasing the productivity of agriculture, forestry and fisheries. Genes from plants, animals and microorganisms that flourish in the forests, fields and seas of the developing world are the strategic raw materials for the commercial development of new pharmaceutical, agricultural and industrial products. Whereas genetic wealth, especially in tropical areas such as rainforests, was once a relatively inaccessible trust fund, it is fast becoming a highly valuable currency. Introduction of modern biotechnologies in the developing world is frequently compared to the Green Revolution. While the Green Revolution involved introducing new varieties of primarily wheat and rice in selected areas, biotechnology has the potential to affect all crops and tree species, as well as fish and livestock, in any corner of the globe. The Green Revolution was introduced to the Third World largely by international institutions, but the "Gene Revolution" is primarily in the hands of the private sector, with transnational corporations being the leading players. With few exceptions, scientific and technical capacity in the biosciences is centred in the industrialized world. As a consequence, biotechnology research, by and large, does not focus on the needs or interests of poor farmers in marginal areas of the world. Emerging biotechnologies have considerable potential to enhance food and agricultural production in the developing world, but they could also add to existing inequities by displacing traditional agricultural products, accelerating genetic erosion and introducing new environmental hazards. Molecular biology is the most powerful tool of biotechnology. In the area known generally as genetic engineering, scientists can transfer genes between unrelated species endowing such "transgenic" plants, animals and microorganisms with properties that they could probably never have acquired in nature. As yet, only a handful of genetically engineered products are available commercially, but hundreds are in the pipeline. Genetic engineers can design crop varieties containing natural insecticidal genes, fish with human growth hormones, and faster growing trees. It must be stressed; however, that genetic engineering consists essentially of mixing and matching genes from different species. It cannot create genetic material, replace lost material or eliminate the need to conserve living resources. Molecular biology is important in characterizing and conserving biodiversity. For example, molecular markers can help establish the extent of diversity within a species and to identify genes of interest to breeders. Such techniques can also help establish priorities for conservation (Peters, 1993).

Biotechnology to Protect Biodiversity

Biotechnology already assists the conservation of plant and animal genetic resources through:

- ✰ New methods for collecting and storing genes (as seed and tissue culture);
- ✰ Detection and elimination of diseases in gene bank collections;
- ✰ Identification of useful genes;
- ✰ Improved techniques for long-term storage;
- ✰ Safer and more efficient distribution of germplasm to users.

Tissue culture is just one example. The technique, which involves growing small pieces of plant tissue or individual cells in culture, provides a fast and efficient way of taking numerous cuttings from a single plant. In many cases, entire plants can be regenerated from a single cell because each cell contains all the necessary genetic information. After selecting a disease-free cutting, for example, scientists can mass-produce copies that are genetically identical. This is the basis of plant cloning, or micropropagation of plants. In gene banks, tissue culture is now used routinely to preserve the genetic information of plants which have seeds that do not store well, are sterile or have poor germination rates. Plant cells maintained on a growth medium in a test-tube replace seeds or plants. Plants stored in this way include sweet potatoes, bananas and plantains, apples, cocoa and many tropical fruits (Malabika Roy Pathak and Mohammad S Abido, 2014).

Biodiversity: Obstacle or...?

Biotechnology contributes to conservation and the sustainable use of biodiversity, but several areas exist where modern biotechnology may hinder development or create serious hardship for rural communities. The economies of developing countries are threatened by biotechnology research that promises to eliminate or displace traditional export commodities, often a primary source of foreign exchange. Current research, for example, focuses on substitutes for tropical oils and fats - ranging from cocoa butter to castor oil. Biosynthesis in the laboratory of high-value ingredients such as vanilla, pyrethrum and rubber could ultimately transfer production out of farmers' fields and into industrial bioreactors. Without ample opportunity to plan and diversify, developing country farmers and their botanical exports may suffer massive displacement, wreaking havoc on already weak economies (Frankel and Brown, 1983).

A New Wave of Genetic Erosion?

Biotechnology may threaten the genetic diversity on which it depends. In the absence of conservation, commercial biotechnology may unleash a new era of genetic erosion. A commercial venture in Chile, for example, can propagate up to 10 million eucalyptus seedlings, all identical clones, in automated nurseries. Similarly, commercial semen and embryo transfer services for domestic animals raise concern about the displacement of traditional livestock breeds. Cloning could accelerate replacement or dilution of indigenous stock by imported breeds, leading to a loss of genetic diversity (Fowler and Mooney, 1990).

Biosafety: A related concern involves the ecological risks of introducing genetically engineered plants into centres of diversity. Transgenic varieties, a good number of them resistant to herbicides, have been produced in more than 40 crop plants. Gene flow to weeds from resistant plants could have far-reaching consequences. The resulting herbicide-tolerant weed could be difficult to control, harming future crop production as well as the surrounding ecosystem (FAO, 1999 and 2000). Will biotechnology firms seeking to penetrate markets in developing countries take into account the risks posed in regions where wild and weedy relatives of major food and industrial crops are found? Will the developing countries have the capacity to monitor and assess the risks?

Opportunity for Development....?

Biotechnologists could develop new varieties and breeds adapted to low-input agriculture or harsh conditions, or improve processing. Biotechnology may help create markets by developing new industrial, medicinal and aromatic crops. Given their richness in biodiversity, several developing countries that have the capabilities, such as Brazil, China and India, could produce new high-value products based on local flora. The congenial agro-ecological settings and availability of relatively cheap labour should be conducive to large-scale production of new high-value crops, enabling such countries to maintain their comparative advantage in these commodities. The use of biotechnology to develop biofertilizers and to detect and control pests and pathogens will be particularly helpful to poor farmers. Such technologies could also bring trading advantages by removing non-tariff barriers arising from the presence of pesticide residues or pest infestation in food commodities that otherwise have an export, and therefore income-generating, potential. The fundamental question posed by biotechnology remains: Who will control the new technologies and benefit from them? FAO is trying to strengthen national capacities to exploit biotechnology for sustainable, low-input agriculture, and to encourage biotechnology research on products/commodities that are important to developing nations. It is also fostering the best uses of biotechnologies to identify and conserve genetic resources. Finally, FAO is developing a Code of Conduct that covers the issues raised above. Global conservation efforts began in the 1950s in response to the rapid erosion of plant and animal genetic resources. A wide range of genetic resources of interest to food and agriculture have now been safeguarded and databases documenting this material continue to be built up. Conservation is not an end in itself, but a means of ensuring that plant and animal genetic resources are available for use by present and future generations.

The two basic approaches to conservation, once the resources are identified and characterized, are *in situ* and *ex situ* methods. *In situ* methods maintain plants and animals in their original habits whereas *ex situ* conservation maintains organisms outside their original habitats in facilities such as gene banks, cell cultures, botanical gardens or zoological parks. The two approaches are not mutually exclusive: different conservation systems can complement each other and ensure against the shortcomings of any one method. With the exception of the small number of varieties and breeds that are widely used, particularly in the developed world, experience shows that diversity is only secure when a variety of conservation strategies is employed (Serageldin, 1999).

Conserving Plant Genetic Resources

Gene banks provide the principal means of storing plant genetic material. This *ex situ* system relies on three methods of storing the material:

Seed banks provide a controlled environment where seeds can be dried to low moisture content and stored at low temperature without losing their viability. Field gene banks such as arboreta, plantations and botanical gardens are useful for species that are difficult or impossible to store as seed, including many perennials, vegetatively propagated crops and tree species.

In vitro methods which conserve plant parts, tissue or cells in a nutrient medium can be used to conserve species that do not readily produce seeds, or where the seeds cannot be dried without damaging them. Under ideal conditions, gene banks provide long-term but not indefinite storage. Both seeds and tissues deteriorate with age and plants must be grown periodically to generate fresh seeds and tissues for continued storage. Unfortunately, even the most sophisticated gene bank cannot always provide adequate security. Large collections of germplasm continue to be lost as a result of technical and financial shortcomings, or natural disasters. Power failures, inadequate documentation and evaluation, or failure to regenerate plants can result in massive losses of stored collections. Earthquakes, floods, social and political unrest also put gene banks at risk. The biggest shortcoming of *ex situ* gene banks is that, once stored, plants are removed from the evolutionary process that they undergo in nature. There is no pressure from changing natural conditions or competition with other species. The ultimate aim of conservation is to maintain a dynamically evolving system. Therefore, while gene banks will continue to be vital to conservation, complementary systems, including *in situ* methods, are likely to assume greater importance in the future (Cook, 2000).

Conserving Animal Genetic Resources

Ex situ conservation of animal genetic resources includes "cryogenic preservation techniques" - the collection and freezing in liquid nitrogen of semen, ova or embryos, or the preservation of DNA segments in frozen blood or other tissues. It also encompasses the captive breeding of wild or domesticated species in zoos or other locations away from their natural environment. *Ex situ* gene banks have similar advantages and disadvantages to those used to hold plant germplasm. *In situ* conservation enables animal populations to continue to evolve and be selected for use in their natural environments. It is particularly important for species or within geographical regions where cryogenic preservation is not well developed or available. *In situ* conservation can be carried out at any level, in any country, without the need for special skills or technology. Properly planned, it can incorporate farmer and industry breeding as well as continued use of the resources (Bruford and Wayne, 1993).

Harmonizing Conservation with Human Needs

In areas where biological resources are being depleted to meet the basic needs of local populations, *in situ* conservation must be reconciled with immediate human needs. Successful conservation depends upon meeting the needs of local people, while ensuring sustainability of the resource. Many approaches are feasible, but these three examples show how, based on scientific expertise and local knowledge and participation, genetic resources can be used and, at the same time, conserved and developed (Thompson *et al.,* 2011).

Who Owns Biodiversity?

International cooperation with respect to biodiversity has been complicated by the efforts of some industrialized countries to extend intellectual property rights to genes, plants, animals and other living organisms, which inevitably leads to restrictions on access to genetic resources. With the advent of genetic engineering, for example, the biotechnology industry has promoted the extension of industrial patenting regimes to living organisms–an approach popularly known as "life patenting". Proponents of patenting argue that it stimulates innovation by rewarding patent holders and enables companies to recoup their research investment. In the 1980s, precedents have been established for extending the concept through "life patenting". As a result, genes, plants, animals and microorganisms–whether discovered in nature or manipulated by genetic engineers–could be rendered the intellectual property of private interests. The patenting of useful genes found in nature is particularly controversial. For farmers and consumers in the developing world it could mean paying royalties on products that are based on their own biological resources and knowledge. Under patent law, a farmer breeding a patented animal and selling its offspring without payment of royalties would be contravening the law. Similarly, it would be illegal for farmers to save seed from a patented variety for replanting (Shands and Stoner, 1997).

Proposals are now being considered at various international flora to:

- Promote the extension of patenting and/or plant breeders' rights in the developing world; and expand the scope of industrial patent law to cover all biotechnology products and processes.
- In promoting mechanisms for rewarding innovators of new biotechnologies, little or no consideration has been given to the impact on the future

conservation and exchange of biological resources. The FAO Commission on Plant Genetic Resources has warned: "... if the patent system is applied universally to living matter, including plants and animals, and their genetic resources, then the principle of unrestricted access will be severely eroded".

- ☆ The danger is that claiming intellectual property rights, without reciprocal benefits and compensation for developing nations, could set up formidable barriers preventing access to genetic resources. In the wake of new intellectual property proposals, developing nations are questioning the concepts of free access and heritage of humankind. They may react by restricting access to germplasm on their territories. Clearly, present proposals could have grave implications for future economic development and world food security.

Crop Plants and their Relatives

The plant genetic diversity used in agriculture–the crops that feed us and their wild relatives–is being lost at an alarming rate. Just nine crops (wheat, rice, maize, barley, sorghum/millet, potato, sweet potato/yam, sugar cane and soybean) account for over 75 percent of the plant kingdom's contribution to human dietary energy. None of the world's staple crops is likely to disappear. Yet they, too, are threatened–not by the loss of a single crop species such as wheat or rice, but by the loss of diversity within species (Arora and Pandey, 1996).

Seeds of Survival

All major food crops, the staple crops grown and consumed by the vast majority of the world's population, have their origins in the tropics and subtropics of Asia, Africa and Latin America. Over the years, farmers selected and domesticated all major food crops on which humankind depends today. Wheat and barley originated in the Near East, for example. Soybeans and rice came from China. Sorghum, yams and coffee came from Africa. Potatoes and tomatoes originated in the Andes of South America, and maize in South and Central America. Crop genetic diversity is still concentrated mainly in regions known as "centres of diversity", and located in the developing world. Farmers in these areas, who still practice traditional agriculture, cultivate local varieties known as "land races" that have been selected over many generations. Closely related species that survive in the wild are known as "wild relatives" of crops. Together, land races and their wild relatives are the richest repositories of crop genetic diversity. Thousands of different and genetically distinct varieties of major food crops owe their existence to millions of years of evolution and to careful selection and nurturing by our farmer ancestors during some 12 000 years of agriculture. This diversity protects the crop and helps it meet the demands of different environments and human needs. Potatoes, for instance, originated in the Andes, but nowadays they can be found growing below sea level behind Dutch dykes or high in the Himalayan mountains. One variety of rice survives on just 60 centimeters of annual rainfall, another float in 7.5 meters of water.

Agriculture's Vanishing Heritage

FAO estimates that since the beginning of this century about 75 percent of the genetic diversity of agricultural crops has been lost. We are becoming increasingly dependent on fewer and fewer crop varieties and, as a result, a rapidly diminishing gene pool. The primary reason is that commercial, uniform varieties are replacing traditional ones–even, and most threateningly, in the centres of diversity. When farmers abandon native land races to plant new varieties, the traditional ones die out. The introduction, beginning in the 1950s, of high-yielding grains developed by international crop breeding institutions led to the Green Revolution. The spread of the new varieties in the developing world was dramatic. By 1990 they covered half of all wheat lands, and more than half of all rice lands–a total of some 115 million ha. This resulted in large increases in yields...but large decreases in crop diversity. The erosion of crop genetic diversity poses a serious threat to food supplies. To maintain pest and disease resistance in major food crops, for instance, or to develop desirable traits such as drought tolerance or improved flavour, plant breeders require fresh infusions of genes from the farms, forests and fields of the developing world. Developing the high-yielding, elite cultivars of modern agriculture depends on a steady stream of new, exotic germplasm. Plant breeders continuously try to develop new varieties to keep one step ahead of thousands of pests and diseases. Without access to traditional land races and their wild relatives, modern agriculture would be seriously endangered (Barbieri, 2013).

Dangers of Genetic Uniformity

Industrialized agriculture favours genetic uniformity. Typically, vast areas are planted to a single, high-yielding variety–a practice known as monoculture–using expensive inputs such as irrigation, fertilizer and pesticides to maximize production. In the process, not only traditional crop varieties, but long-established farming ecosystems are obliterated. Genetic uniformity invites disaster because it makes a crop vulnerable to attack–a pest or disease that strikes one plant quickly spreads throughout the crop. The Irish Potato Famine of the 1840s is a dramatic example of the dangers of genetic uniformity. None of the few varieties of the New World potato introduced into Europe in the 1500s were resistant to a potato blight that struck Ireland in the 1840s. The potato crop was wiped out. Over a million people died in the famine and a million more emigrated to the New World. More recently, in 1970, genetic uniformity left the United States maize crop vulnerable to a blight that destroyed almost $1 000 million worth of maize and reduced yields by as much as 50 percent. Over 80 percent of the commercial maize varieties grown in the United States at that time were susceptible to the virulent disease, southern leaf blight. Resistance to the blight was eventually found in an African maize variety called Mayorbella. A major catastrophe was averted by incorporating this resistance into commercial varieties.

The Value of Crop Genetic Diversity

The value of genetic diversity to modern plant breeding is enormous. The United States Government estimates that a 1 percent gain in crop productivity means a $1 000 million benefit to the American economy. Italian scientists calculate that the benefits of exotic germplasm for a single crop, durum wheat, amount to $300 million per year. Not only cultivated species but also the genes from wild relatives are enormously valuable. Between 1976 and 1980, wild species contributed an estimated $340 million per year in yield and disease resistance to the farm economy of the United States.

Stunted Rice: A Wild Plant to the Rescue

During the 1970s the grassy-stunt virus devastated rice fields from India to Indonesia, endangering the world's single most important food crop. After a four-year search which screened over 17 000 cultivated and wild rice samples, disease resistance was found. Only one population of the species *Oryza nivara*, growing wild near Gonda in Uttar Pradesh, was found to have a single gene for resistance to grassy-stunt virus strain 1. Today, resistant rice hybrids containing the wild Indian gene are grown across 110 000 km2 of Asian rice fields. In the developing world, crop genetic diversity enables farmers to select crops suited to ecological needs and cultural traditions. Without this diversity, options for long-term sustainability are lost. This is particularly true in marginal areas with highly varied environments. The variety to a large extent determines the need for fertilizers, pesticides and irrigation. Communities that lose traditional varieties, adapted to local needs and conditions over centuries, risk becoming dependent on external sources of seeds and the inputs needed to grow and protect them. Without an agricultural system in harmony with a community and its environment, self-reliance in agriculture is impossible. To feed an increasing world population, all available genetic resources, including wild relatives, will need to be tapped. Modern plant breeding as well as new biotechnologies offer the potential to exploit little-known plant species as sources of food, and to enhance the qualities of those plants that are underutilized–especially traditional plants of special significance to poor people, such as local grains, legumes, oilseeds, fruits and vegetables.

Traditional food crops, often grown by rural families to see them through the "hungry season" just prior to harvest, offer many advantages. Many of them are drought resistant, can be grown without expensive inputs and have good storage qualities. For many developing nations, self-reliance in food production will depend on low-input agriculture in poor production environments. The capacity to grow varieties, particularly those resistant to pests and diseases and adapted to marginal lands, is vital for sustainable agriculture and food security.

Geopolitics of Plant Genetic Resources

Historically, scientists from the industrialized countries have ventured southwards in search of exotic plants for plant breeding. Seeds found in tropical centres of diversity have been freely collected and later deployed in plant breeding. As a result, much of the collected diversity of Third World origin has come to be stored in the northern hemisphere or in gene banks established by developed countries. The issue of control, ownership and access to plant genetic diversity has come to the fore over the past two decades. Plant breeding in the industrialized world has become increasingly commercialized and is now dominated by transnational seed and agrochemical corporations. To promote innovation and to enable breeders to recoup their research investment, many governments in the industrialized world have adopted a system of "plant breeders' rights". This gives patent-like protection to breeders with limited monopoly rights over the production, marketing and sale of their varieties for a period of up to 20 years. The disparity between unrestricted access to genetic resources, including farmers' land races, and the existence of proprietary rights such as "breeders' rights" on improved varieties has fuelled intense debate over the inequity in the flow of germplasm from the developing world to the industrialized world. At the United Nations, representatives from the developing world ask: Why are patented seeds, originally from developing countries, bringing profits to seed companies in the industrialized countries without corresponding compensation for the developing world? What compensation will be made to those who have tended and nurtured the world's crop genetic diversity and continue to conserve and make it available today?

Promoting the Use and Conservation of Plant Genetic Resources

The farmer uses plant genetic resources as seeds or vegetatively propagated material; they are often the one input that farmers can produce for themselves. FAO assistance includes projects for the production and use of good-quality seed, training and guidance in propagation and multiplication, quality control, and processing, storage and distribution of improved seed. FAO provides samples and information to research centres, scientists and field projects for use in crop introduction, evaluation and breeding.

In Conclusion Biodiversity to Nurture People

Biodiversity provides the raw materials, combinations of genes, that produce the plant varieties and animal breeds upon which agriculture depends. Thousands of different and genetically unique varieties of crops and animal breeds owe their existence to 3 000 million years of natural biological evolution and to careful selection and nurturing by our farming and herding ancestors during 12 000 or so years of agriculture.

Whether they are used in traditional farming systems, conventional or modern breeding or genetic engineering, the genetic resources of plants and animals are a global asset of inestimable value to humankind. As genetic diversity erodes, our capacity to maintain and enhance crop forest and livestock productivity decreases along with the ability to respond to changing conditions. Genetic resources hold the key to increasing food security and improving the human condition.

REFERENCES

Agarwal, N.K., Singh, G. and Singh, H., 2011. Present status of Icthyofaunal diversity of Garhwal Himalayan river Bhilangna and its tributaries with reference to changing environment. Environment Conservation Journal, 12(3): 101-108.

Alabi, K.A. 1992. Traditional Forest Conservation in Practice: A Case Study of Ikire-Ile In Ola-Oluwa Local Government. M. Sc. Degree Dissertation. Department of forest resources management. University of Ibadan, Nigeria.

Arora, R.K. and Pandey, A. 1996. Wild Edible Plants of India: Diversity, Conservation and Use. New Delhi: [Indian Council of Agricultural Research] National Bureau of Plant Genetic Resources.

Ayoade, A. A., Agarwal, N.K and Chandola-Saklani, A., 2009. Changes in Physico-chemical Features and Plankton of Two Regulated High Altitude Rivers, Garhwal Himalaya, India. European Journal of Scientific Research, 27 (1): 77-92.

Babu SC (2000). Rural nutrition interventions with indigenous plant foods: a case study of vitamin A deficiency in Malawi. Biotechnol. Agron. Soc. Environ., 4(3): 169–179.

Barbieri, C. 2013. Assessing the sustainability of agritourism in the US: A comparison between agritourism and other farm entrepreneurial ventures. Journal of Sustainable Tourism, 21(2), 252– 270.

Boyer JS. 1982. Plant productivity and environment. Science, 218: 443–448.

Bruford, M.W. and Wyne, R.K. 1993. Microsatellites and their application to population genetic studies. Curr.Opin.Genet. Dev. 3: 939-943.

Cook, R. James. 2000. "Science-Based Risk Assessment for the Approval and Use of Plants in Agricultural and other Environments." In G.J. Persley and M.M. Lantin, eds., Agricultural Biotechnology and the Poor: Proceedings of an International Conference, Washington, D.C., 21-22.

Dalal-Clayton, B. and Bass, S. 2002. Sustainable Development Strategies: A Resource Book. Organization for Economic Co-operation and Development (OECD), Paris and United Nations Development Programme (UNDP), New York. 358.

FAO. 1999. Information note on biosafety, presented at the Thirtieth Session of FAO Conferences, page 7.

FAO. 2000. Agriculture Towards 2015/2030, a technical interim report.

Field, R., Hawkins, B. A., Cornell, H.V., Currie, D.J., DinizFilho, J., Alexandre, F., Guégan, Jean-François; Kaufman, D.M., Kerr, J.T., Mittelbach, G.G., Oberdorff, T., O'Brien, E. M. and Turner, J. R. G., 2009. Spatial species-richness gradients across scales: a meta-analysis. Journal of Biogeography, 36 (1): 1365-2699.

Fowler, C. and P. Mooney. 1990. Shattering: Food, politics, and the loss of genetic diversity. University of Arizona Press, Tucson, AZ.

Frankel, O. H. and Brown, O. H.D. 1983. Current plant genetic resources-A critical appraisal. Oxford and IBH Publishing, New Delhi.

Gaston, K.J., 2000. Global patterns in biodiversity. Nature, 405 (6783): 220–227.

Harris MA, Mohammed S. 2003. Relying on nature: wild foods in Northern Nigeria. Ambio, 32(1): 24–29

John Dixon and Aidan Gulliver with David Gibbon. 2001. Farming Systems and Poverty Improving Farmers' Livelihoods In A Changing World. FAO and World Bank Rome and Washington DC.

Ludlow MM, Muchow RC. 1990. A critical evaluation of the traits for improving crop yield in water limited environments. Adv. Agron. 43: 107–153.

Malabika Roy Pathak* and Mohammad S Abido. 2014. The role of biotechnology in the conservation of biodiversity. Journal of Experimental Biology and Agricultural Sciences, 2(4): 353-363.

Mora, C., Tittensor, D.P., Adl, S., Simpson, A.G. and Worm, B., 2011. How many species are there on Earth and in the ocean? PLOS Biology. 9(8).

Peters, P. 1993. Biotechnology: a guide to genetic engineering: Wm. C. Brown Publishers.

Rabosky, D.L., 2009. Ecological limits and diversification rate: alternative paradigms to explain the variation in species richness among clades and regions. Ecology Letters, 12 (8): 735–743.

Serageldin, I. 1999. Biotechnology and Food Security in the 21st Century. Science 285: 387-389.

Shands HL, Stoner AK (1997) Agricultural germplasm and global contributions. In KE Hoaglund, AY Rossman, eds, Beltsville Symposia in Agricultural Research: Global Genetic Resources: Access, Ownership, and Intellectual Property Rights. Association of Systematics Collections, Washington, DC, 97–106.

Shaw, D. 1998. Risks or remedies ? Safety aspects of herbal remedies. J. Roy. Soc. Med., 91, 294–296.

Thompson ID, Okabe K, Tylianakis JM, Kumar P, Brockerhoff EG, Schellhorn NA, Parrotta JA, Nasi R. 2011. Forest biodiversity and the delivery of ecosystem goods and services: Translating science into policy. BioScience 61: 972–981.

Tiisekwa BPM, Ndabikunze BK, Samson G, Juma M (2004). Suitability of some indigenous tree fruits for manufacturing juices and jams in Tanzania. In: Rao MR, Kwesiga FR (eds). Agroforestry impacts on livelihoods in southern Africa: Putting research into practice. Proceedings of the Regional Agroforestry Conference held in Warmbaths, South Africa 20–24 May, 2002. World Agroforestry Centre (ICRAF), Nairobi, Kenya, pp. 331–335.

Chapter 5

Propagation and Conservation Strategies in Mulberry (*Morus* spp.)

Z.I. Buhroo, Shahina A. Nagoo, Sabiha Ashraf, S.A. Wani and Masarat Bashir*

College of Temperate Sericulture-Mirgund, Sher-e-Kashmir University of Agricultural Sciences and Technology of Kashmir Srinagar – 190 025, J&K
**Corresponding author e-mail: zafar.buhroo1985@gmail.com*

Mulberry (Morus L.) is an important tree crop that provides sustainable economic and environmental benefits to a large number of people who live in rural and suburban areas of Asian countries especially of China and India. Mulberry trees (genus Morus) are small to medium-sized trees belonging to the Moraceae family. Mulberries are native to temperate Asia and North America, and many varieties are grown as ornamentals and for their fruits. Mulberry plants are also important as food for silkworms. Mulberry plants are also essential to silkworms as a source of food. it is believed that more than 68 species exist under the genus Morus and out them only a few species such as M. alba, M. bombycis, M. indica, M. latifolia, M. multicaulis (for foliage) and M. nigra (for fruit) are cultivated. The remaining species along with many landraces of the cultivated species are considered as wild, therefore, they have been mostly neglected. Cuttings and saplings can also be used to expand plantations. Plant propagation is the method of growing new plants from various sources, including seeds, cuttings, and other plant parts. The numerous propagation techniques used in mulberry cultivation are discussed in this chapter. countries across the globe have been adopting both conventional and modern biotechnological methods to collect, characterize, conserve and utilize large amount of genetic resources in their crop improvement programs. This, review is, thus, undertaken to give an overview of the role of biotechnology on exploration, characterization, and conservation of wild germplasm for crop improvement in mulberry.

***Keywords**: Conservation, Morus, Mulberry, Propagation.*

Introduction

Mulberry (Morus; Moraceae), is a tree crop being cultivated widely in Asian countries to feed the monophagous Silkworm (*Bombyx mori* L). Since mulberry leaf production cost along covers more than 60 per cent of the total production cost of silkworm cocoons, mulberry cultivation is considered as one of the vital components of sericulture activities (Das *et al.,* 1969). Generally, the plant has three main species ostensibly named after the fruit color such as white (*Morus alba*), red (*M. rubra*) and black mulberry (*M. nigra*) with numerous cultivars but most of these mulberry varieties were developed from a few species such as *M. alba, M. atropupurea, M.bombycis, M. indica, M. latifolia* and *M. multicaulis* though more than 68 species have been reported from the genus Morus (Datta, *et al.,* 2000). The major reasons for the lack of utilization of other species are the poor leaf quality due to the, coarseness of the leaf, low moisture content and its retention after leaf harvest, low protein content, and the low leaf yield (Das *et al.,* 1984). Thus, it is imperative to develop qualitatively and qualitatively superior mulberry through identification of better parents and their utilization in breeding. Mulberry is a woody deciduous perennial plant with a rapid growth rate. Its root system is comprehensive. Easy, alternate, stipulate, petiolate, whole, or lobed leaves are found on this plant. Plants are generally dioecious. Inflorescence is catkin with pendent or drooping peduncle bearing unisexual flowers. Inflorescence is always auxiliary. Catkins produced by males are normally longer than those produced by females. Male flowers are loosely arranged and after shedding the pollen, the inflorescence dries and falls off. These are four persistent parianth lobes and four stamens implexed in bud. Female inflorescence is usually short and the flowers are very compactly arranged. There are four persistent parianth lobes. The stigma is bifid and the ovary is one-celled. Wind is the most important pollinator in mulberry. Mulberry fruit is mostly violet black in colour and is called sorosis.

Plant propagation can be characterised as the multiplication or reproduction of plants in simple terms. Plant propagation methods and procedures have evolved as a result of the commercialization of crops. Each technique has its own set of advantages and disadvantages. Each plant responds differently to different methods of propagation. Plant propagation is the skillful integration of science and art. It aids in the conservation of plant stocks and the preservation of endangered (extinct) animals. Further, improvement of productivity traits in mulberry plays a vital role in the progress of sericulture industry (Biasiolo *et al.,* 2004; Dandin and Naik 2004), while improvement of mulberry tree through conventional breeding has substantially contributed to the success of sericulture industry. Further, the application of biotechnology techniques in mulberry crop improvement holds a great promise especially in those areas where conventional research has not achieved the desired success (Dandin *et al* 2004; Guha *et al.,* 2010).

Methods of Propagation

Mulberry can be propagated by seeds or vegetative parts (Figure 5.1), however vegetative method is common, easy and cheap. Further it is advantageous to get desired characters like maintenance of particular characters of the plant, speed in raising large number of saplings, adaptability to a particular habitat, resistance to pests and diseases and produce viable seeds. Due to bi parental origin a seed cannot produce a true type of plant. Different vegetative propagation methods are followed in different countries depending on the environmental conditions and soil nature. In India mulberry is propagated through common and popular method-cuttings in multivoltine areas like Karnataka and West Bengal. In other parts root grafting is followed where exotic varieties cannot be raised by cuttings.

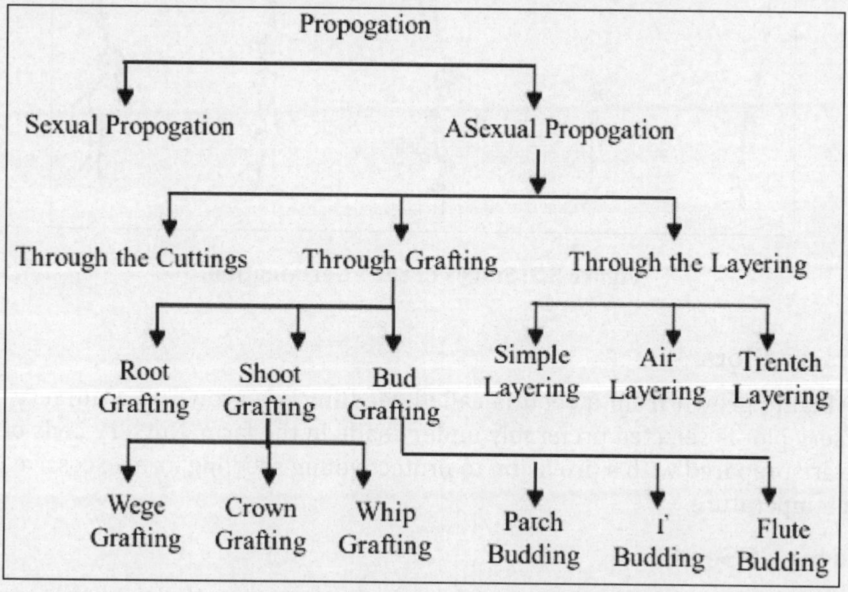

Figure 5.1: Methods of Propagation.

Sexual Propagation

In this mulberry is propagated through seeds. This method is very easy and cheap. Sexual propagation is mainly to bring about a varied population for the purpose of selection and hybridization. The seeds are used to obtain stock material for grafting. It is also suitable for large-scale multiplication. But long gestation period to provide learner for silkworm rearing and desirable traits of improved cultivars cannot be perpetuated. The different stages that are involved in the sexual propagation (Figure 5.2) are as under:

Seed Collection

Seeds are collected from ripe fruits in March-April in tropical region and May-June in temperate region. There is no dormancy in mulberry seeds. The mulberry seeds are small and about 300,000 seeds are required to fill one litre volume.

Figure 5.2: Stages of Seed Germination.

Nursery Plots

Plants grown from a seed is called seedling are grown in a nursery. The nursery plot is selected preferably under shade in the farm. Nursery beds of 0.9, size are prepared with a provision to protect young seedling form excessive light and temperature.

Sowing of Seeds

The mulberry seeds are soaked for a day in water to soften the hard testa for easy and successful germination. Seeds are broadcast or sown in holes in nursery beds. The seeds should not sow deeper than 2.5 cm to avoid delay in germination or total failure.

Seed Germination

The environmental conditions *i.e.*, temperature, light play vital role in seed germination. Therefore seed beds are covered with dams made of bamboo strips or palm or coconut leave to protect from severe sun.

Asexual Propagation

It is the most popular method adopted in moriculture. It favours to utilize vegetative parts of a plant to grown them into a individual plant. This propagation method is very easy though it is technical. There are different methods to utilize branches, buds, roots in this propagation.

Cuttings

Most of the tropical varieties of mulberry can be easily propagated on large scale through cuttings. It is easiest, common and popular method in India. But this method is restricted to acclimatized local varieties. The plants are selected on the basis of nutritious leaf, high yield, quick growth resistance to diseases and insect pests and drought resistant varieties. This propagation is easy, cheap and quick. Mulberry being a cross pollinated plant, the improved traits of the cultivar are retained by this method.

The branch must have 10-12 mm in diameter. The cutting of, 18-20cm long with three or four active buds are prepared with a sharp implement so that neither the bark nor the wood portion is damaged. The ends must be slant (45 deg), without damaging the bark or splitting base.

(Cutting size 18 -20 cm long, 3-4 buds - Irrigated farm)

(Cutting size 20-24 cm. long, 5-6 buds - Rain fed farm)

The roots develop from the basal end of the cuttings. The roots develop endogenously. The root initials developed initially finally produce root primordium so as to form root tip, which grows outward and emerges at right angles to the stem. IBA treatment gave generally higher rooting success between 56.7 per cent and 89.3 per cent (Ozkan *et al.*, 1996). Simultaneously callus tissue develops at the basal end of the cutting and becomes stem of the future plant.

Root hormones are used to certain mulberry varieties which don't produce roots from cuttings. This is followed in temperate regions. They are Indole 3-acetic acid (IAA), Indole 3-butyric acid (IBA), Naphthalene acetic acid (NAA), 2,4 – dichlorophenoxy acetic (2, 4-D), rootone, senadix. The chemicals act like auxins and promote root development.

Grafting

It is a technical joining of a branch of one plant into a rooted plant in such a way that they unit with an organic union formed in between them, and finally grow as one plant. It is practiced where the plant cannot be propagated through cutting, because of poor rooting.

Scion - The plant part of the graft combination which is to become the upper portion of the shoot system of the new plant.

Stock - The lower plant part with root base, which supports and supplies nutrients to scion, to grow as new plant. Thus the stock must be a local, handy variety and scion is selected from the desired or indigenous variety.

Basic principles of grafting are:

a) Compatibility of stock and scion
b) Maximum cambial contact between scion and stock
c) Local adaptability of stock and superiority of scion.

Depending upon the type of material used, its nature age and portion of the stock, grafts are of the following types: wedge and crown grafting, whip, root, bud grafting.

Shoot Grafting

The scion is inserted into the stock stem is called shoot or stem grafting. It is of many types.

Wedge Grafting

It is a method to remove the old plant. The plant is pruned at a convenient height and a 'V' shaped incision in made at the cut surface. The basal region of scions cut obliquely to fit in the incision made on the stock plant.

Crown Grafting

It is a method where more than one scion is inserted into the stock to get a bushy plant. Crown grafting is generally used in propagation of white mulberry (*M. alba*) plants for silkworm growing (Tingzing *et al.*, 1988).

Whip Grafting

A slope cut is given to a length of 3.5 to cm to the stock of 1.2 to 2.5 cm thick. The scion of same size is cut in similar process so as to fit in the stock, and inserted into the stock and tied firmly with soft fibre. The site of join is sealed with grafting wax or clay.

Root Grafting

The root is used as a stock instead of shoot. Roots of 0.6 to 2.5 cm thick are selected from a local, one year old seedling and cut into pieces of 5 to 7.5 cm long. The top end is cut obliquely. The scion of 8 to 10cm long, little thinner than stock and with two or three buds is prepared (Figure 5.3). The basal end should be oblique with the end tapering to provide a stiff tongue. The scion is inserted at the pointed end of the stock between wood and bark. The cut ends of both stock and scion should be in the same direction thus cambial layers are in close proximity to each other. This method is simple, highly efficient, easy, quick economical and good for root development. The percentage of success is more and an individual can prepare 800-1000 grafts in eight hours. A two year old seedling given five to six stocks and all can be used as grafts to get five to six plants, which minimizes cost of labour. But it is necessary to remove the latent buds completely.

Bud Grafting or Budding

It is a better method when the scion material is scarce. In this grafting only one vegetative bud with a piece of bark is used as scion material,thus called as bud grafting, which has all the advantages of grafting. It is much practiced in temperate regions of the country. The removed bud piece contains periderm, cortex and phloem which are laid against the exposed xylem of the stock (Figure 5.4). There

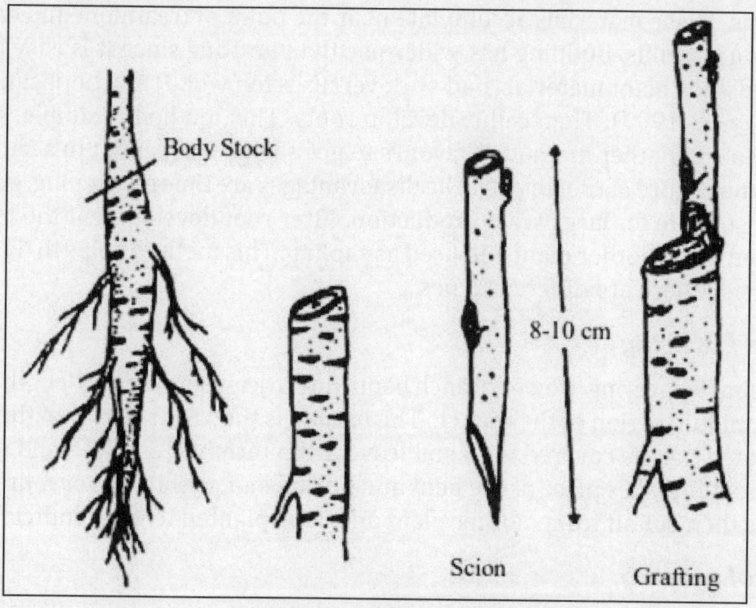

Figure 5.3: Root Grafting.

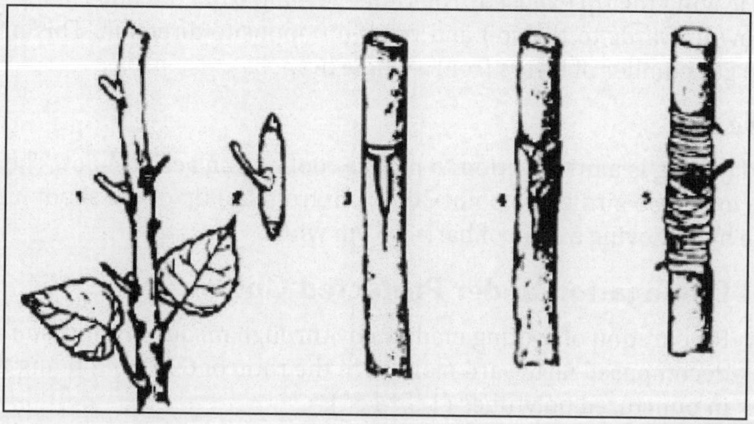

Figure 5.4: Budding.

are various types of budding depending upon the types of the bud. They are 'T' budding, inverted 'T' budding, patch budding, flute budding, chip budding and ring budding.

Layering

It is a process of development of roots on a stem while it is still attached to the parent plant. The rooted stem is detached to grow as a new plant and called as a layer. Root formation on the stem during layering is stimulated by various treatments which cause the interruption in the downward translocation of organic material (carbohydrates, auxins and other growth regulators) from the leaves and shoot tips. These materials accumulate near the point of treatment and cause the formation of callus. Budding has wider use than grafting since it is easy to apply and needs less plant material (bud stick versus scion wood) and application time (Kester *et al.*, 1997). Then callus develop roots. This method is simple, safe and cheap than any other method. It favours to get a large sized plant in a short time. There is no chance of root drying. The disadvantages are time consuming, expensive and not suitable for large scale production. After root development the branch is separated from mother plant and used as sapling. This method helps to fill up gaps in the field. There are different types.

Simple Layering

In simple layering a lower branch bent and bark is removed to a length of 2.5 to 3 cm in middle region of the branch. The branch is then bent slowly to the ground and treated part is covered with soil leaving the terminal end free and exposed. Root initiation takes place at the bent and buried end, within two or four months, which is then cut off from mother plant and transplanted to grow individually.

Trench Layering

In trench layering a branch of growing plant is bent and covered with soil and manure leaving the tip exposed. The roots develop from the buds along with the stem grow upwards to the soil and roots on opposite direction. This process is better to get number of layers from a single branch.

Air Layering

Air layering is a propagation to induce roots on an aerial shoot. The method involves making a girdle at a point 20-40 cm from the tip of the shoot just below the node by removing a strip of bark 2-5 cm wide.

Clonal Propagation under Protected Cultivation

- ✰ Preparation of rooting medium by through mixing of sand, soil and well decomposed farm yard manure in the ratio of 6:2:1 and filling the same in punctured polytubes (11"×4.5").
- ✰ Preparation of 1.5-2.5 cm thick and 15-20 cm long cuttings having 3-4 active buds, from 9-12 old month old healthy shoots of mulberry (Figure 5.5).
- ✰ Treatment of cuttings with 0.02 per cent of Bavistin or Dithane M-45 by dipping them for about 10 minutes to protect the cuttings from any fungal infection. Planting one cutting per polytube during last week of March to first week of April, leaving only one bud exposed followed by light irrigation.
- ✰ Maintaining optimum optimum hygrothermic conditions (temperature 30±5° C and relative humidity of 80±5 per cent). In case the day temperature in polyhouse rises beyond this range, remove the polythene

from the sides and keep the doors open till teh temperature reaches the optimum.

Figure 5.5: Raising of Mulberry Cuttings under Polyhouse Technology.

- ☆ Irrigation should be given and weeding done as per need.
- ☆ The saplings after establishment of root are transplanted to nursery. The nursery already prepared by digging (2-3 times), levelling and mixing 8-10mt of well decomposed FYM is divided into beds of convenient size. The saplings are taken from the polyhouse and planted at least 9×9 inches along with the rooting medium by cutting the polytubes with a blade during the second fortnight of June.
- ☆ The space between sapling lines is covered with black polythene as mulch.
- ☆ Water the nursery at least once in a fortnight or as per need till October last.
- ☆ During 2nd year (after 15th March till Oct), water the plants at least once in a fortnight.
- ☆ Remove the side branches regularly leaving only the main branch to grow.
- ☆ Apply chemical fertilizer in the form of NPK@50:25:5Kg/ha/year.
- ☆ To avoid rodent damage, rodenticide should be applied to beds during November- December.
- ☆ The saplings are fit for supply in the last week of March taking exactly 2 years from cutting plantation to sapling supply.

Tissue Culture in Mulberry

Tissue culture technique in mulberry has developed and ramified into different areas such as micropropagation, callus culture, organogenesis, and screening of genotypes for stress tolerance, induction of polyploids, cryopreservation, and transgenesis.

Micropropagation

Although propagation through stem cuttings is easy and economical, its success depends on a number of factors as mentioned above. Furthermore, newly developed mulberry varieties cannot be immediately propagated through stem cuttings into large areas. Additional issues with growing saplings in a nursery, such as care and management costs for 6-12 months 9 Kapur *et al.*, 2001), seasonal influence (no propagation can be done during the winter period), and the loss of vigour of plants grown from stem cuttings compared to seedlings (Dapat *et al.*, 1987; Zaman *et al.*, 1997) necessitate a viable alternative with which mulberry can be propagated easily and cost-effectively. Since a large number of clones can be generated in a relatively short time and space, micropropagation is an alternative method for the rapid and cost-effective multiplication of mulberry. It has been demonstrated for the first time (Ohyama, 1970) that complete plants could be regenerated if explants from axillary buds of *M. alba,* are cultured on (Murashige and Skoog 1962) medium supplemented with growth regulators.

Effect of Media on Shoot Proliferation

Media composition is another key factor that affects the success of micropropagation of any plant. Therefore, much effort has been made to standardize the culture medium. The basal medium found best suited for mulberry shoot multiplication is full-strength MS (Murashige and Skoog 1962) medium, although other media like AE (Arnold *et al.*, 1981), B5 (Gamborg *et al.*, 1968) and woody plant medium (Lloyd and McCown 1980) also showed varying degrees of success (Kim *et al.*, 1985).

Effect of Media on Rooting

Production of plantlets with profuse rooting *in vitro* is important for successful field establishment of *in vitro* regenerated plants (Murashige and Skoog 1962). Therefore, the effect of genotype, media composition and PGRs on root growth and development in mulberry was investigated in detail. It is reported that 95 per cent success in rooting from *M. alba* cultivar 'S1' in half-strength MS medium supplemented with 0.5 mg L^{-1} NAA. Recently, several auxins such as NAA, indole-3-acetic acid (IAA), IBA and 2,4-D were used singly or in combination to induce rooting from *in vitro* raised shoots of many cultivars.

Hardening and Transfer of Plants into the Field

Hardening of plants developed *in vitro* is one of the critical steps in the field establishment of micro propagated plants. Achievement of uniform plant growth and high survival rate not only demand good greenhouse conditions, but also modifications of the internal micro-climate to match the local environment (Mondal *et al.*, 2004). The plants are transferred to normal growing conditions in the greenhouse and kept there for 6 months. After 6 months growth the plants were transferred to the field and planted at 60 cm × 60 cm spacing.

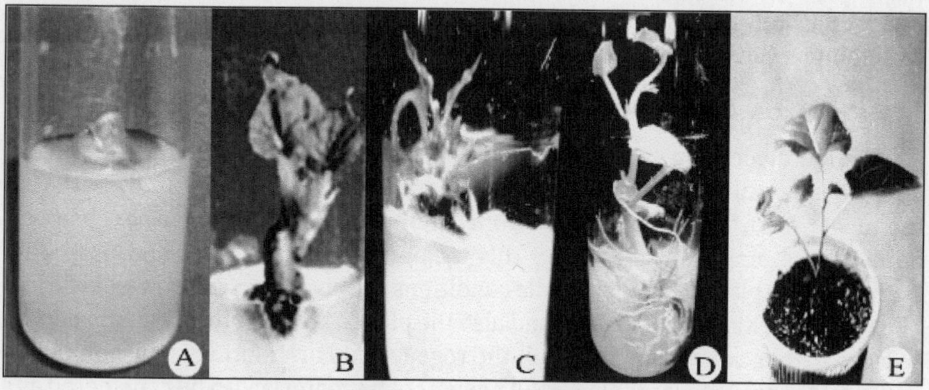

Figure 5.6: Micropropagation in Mulberry.
A: Nodal explant, B: Shoot from the explant, C: Multiple shoots growing from explant, D: Rooting of the micropropagated plant, E: Hardening of the micropropagated plant.

Field Performance of Micro-propagated Plants

Although a number of reports are available on micropropagation of mulberry, only one or two studies have been conducted to assess the field performance of micro propagated plants. It has been demonstrated that using three cultivars *in vitro* raised mulberry plants were more vigorous in growth than those raised from stem cuttings (Zaman *et al.*, 1997). However, unlike morphological traits, the biochemical characters did not show any significant differences.

Organogenesis in Mulberry

Plant regeneration from somatic cells has a high impact in woody tree improvement as it facilitates the development of transgenic plants, isolation of somaclonal variants, development of stress-tolerant plants through cell line selection and multiplication of desirable genotypes (Bajaj YPS, 1986). Therefore, attempts have been made in mulberry to develop protocols for direct plant regeneration from explants like leaves, cotyledons, and embryos and also for indirect plant development via callus culture. Direct organogenesis from explants has the advantage of inducing the least genetic variation among regenerated plants, *i.e.* clonal propagation whereas organogenesis from callus has the advantage of higher plant productivity; moreover, plants developed from callus serve as a significant source of variation for isolation of somaclonal variants with desirable traits.

Biotechnological Tools for Conservation of Mulberry Genetic Resources

With the advent of biotechnology, the conservation strategies for plant genetic resources undergone significant changes as it facilitated conservation of genetic resources more economically with fewer genetic changes, especially for those which are propagated through vegetative mans. The following advanced conservation technologies have been adopted for mulberry.

In vitro Conservation

Mulberry being a vegetatively propagated plant is conserved under *in vitro* conditions by culturing shot-tips and axillary buds. Somatic embryogenesis and organogenesis from callus obtained from different explants could also be conserved. Some of the major advantages of this technique is to conserve and propagate pathogen free plants as under sterile conditions the plants multiplied would be free of pathogens (Kartha KK., 1986) and also the plants can be multiplied irrespective of the weather conditions. Since all the regenerants originating from cloning are identical to the mother plant, a small piece of plant material can be considered as a germplasm and it can be used for conservation. Two types of conservation strategies have been used mulberry such as the slow growth method and the cryopreservation method. In slow growth method, shoot tips or axillary buds are cultured under conditions which keep the plant alive but retard the growth significantly. In general, shoot tips and nodal segments of juvenile shoots are used for micropropagation in mulberry as they limits the incidence of contaminations (Vijayan *et al.,* 2015) besides, Embryos, hypocotyls, cotyledons and leaf tips are also be used as explants for micropropagation (Kim *et al.,* 2015; Thomas T. D, 2003). The most widely used medium is the MS medium (Murashigeand Skoog1962) and other media like AE (Von and Eriksson, 1981), B5 (Gamborg, 1968) and woody plant medium (WPM) (Lloyd andCown, 1980) and growth hormones such as 2mg L-1 BAP, NAA, IAA, IBA, 2,4-D alone or in various combinations (Chitra and Padmaja, 1999). However, for long term preservation of genetic materials slow growth method is not useful as it required repeated regeneration and may suffer from genetic changes and contaminations from pathogens (Ashmore, 1997). Thus for long term preservation cryopreservation methods are used.

Cryopreservation of Dormant Buds

Cryopreservation involves storage of plant material at ultra-low temperatures in liquid nitrogen (-196°C). At this temperature, cell division and metabolic activities remain suspended and the material can be stored without changes for long periods. It also takes very little space for storage of large number of genetic resources. In mulberry the ideal plant part for cryopreservation was found to be wintor buds, though embryonic axes, pollen, synthetic seeds have also been used (Niino, 1995, Niino and Sakai, 1992*a,b,c,d*). There two type of cryopreservation techniques suitable for mulberry. Under the conventional method, cryopreservation for material is achieved by slow cooling down of the material at a controlled rate (usually 0.1-4°C/min) down to about –40°C, followed by rapid immersion of samples in liquid nitrogen. It is operationally complex and requires sophisticated and expensive programmable freezers. On the other hand, the new method called vitrification procedure, the cell dehydration is achieved by subjecting the samples to physical or osmotic dehydration. This is followed by ultra-rapid freezing which results in vitrification of intracellular solutes, *i.e.* formation of an amorphous glassy structure without occurrence of ice crystals, which are detrimental to cellular structural integrity. These techniques are less complex and do not require a programmable freezer, hence are suited for use in any laboratory with basic facilities for tissue culture (Rao, *et al.*, 2007) preserved 908 mulberry germplasm accessions belonging to *M. indica, M. alba, M. latfolia, M. cathayana, M. laevigata, M. nigra, M. australis, M. bombycis, M. sinensis, M multicaulis* and *M. rotundioba.*

DNA Banking

Another method of genetic resource preservation is the DNA banking wherein the genetic information encoded in the genome of the plant conserved for long by isolating the nucleic acid and preserving it in -80°C in deep freezers. Mulberry genomic DNA can be extracted from leaf samples using different routinely used techniques (Vijayan., 2004). Although the pure DNA can be stored in several ways, the most commonly used method is dissolving the DNA in Tris TE buffer (10 mM Tris- HCl and 1.0 M EDTA) and storing at -80°C in deep freezers. The purified DNA can also be stored in alcohol indefinitely as DNA is highly stable and it is estimated that a fully hydrated DNA can be kept at room temperature about 10,000 years to depolymerise into small fragments (Mandal, 1999). Lyophilized DNA (dried) can also be stored indefinitely without any damage. Since, DNA can be stored in small vials, it is presumed that the genetic resources of all the plant species in the world can be stored in a single building (Ford, 1990).An alternative to preservation of isolated DNA is direct storage of cells and tissues under low temperature (-80°C). This method has the advantage of providing a continuous supply of materials for extraction of DNA and also provides materials for biochemical and molecular studies. DNA banking is considered as the most economical way of conserving the plant genetic resources

Conclusion

The efficient utilization of genetic resources depends largely on appropriate characterization, evaluation, documentation and conservation of the mulberry genetic resources using both by conventional and biotechnological approaches. Mulberry, as a multi-purpose tree, has the ability to contribute significantly to global agriculture. It's a highly adaptable plant that can play a variety of roles in small-scale agricultural production. There is much scope to expand and diversify its applications. This species has a lot of genetic diversity and can adapt to a lot of different soil and climatic conditions. Conservation of plant genetic resources is rapidly becoming increasingly important, especially for sustainable agriculture to feed the needs of ever-increasing populations. Although seeds of sexually propagated plants can be conserved easily, conservation of vegetatively propagated plants requires species techniques and attention. Traditional conservation strategies like *In situ* and *ex situ* conservation should be complemented with modern techniques like cryopreservation and DNA banking as *ex situ* conservation of mulberry entails huge investment in the form of labour and space. Thus, concerted efforts can be made to integrate all these techniques to conserve precious genetic resources of this very important tree crop of Asia. Sericulture in general and mulberry cultivation in particular got much benefit from modern biotechnological tools in its conservation and productivity improvement.

REFERENCES

Ashmore SE. Status report on the development and application of in vitro techniques for the conservation and use of plant geneticresources. 1997;67(2): 152-170.

Bajaj YPS (1986) Biotechnology of tree improvement for rapid propagation and biomass energy production. In: YPS Bajaj (Ed) Biotechnology in Agriculture and Forestry (Vol 1) Trees, Springer Verlag, Berlin, pp 1-22.

Bapat VA, Mhatre M, Rao PS (1987) Propagation of Morus indica L. (mulberry) by encapsulated shoot buds. Plant Cell Reports 6, 393-395.

Biasiolo M, Canal DAMT, Tornadore N (2004) Micromorphological characterization of ten mulberry cultivars (Morus spp.). Econ Bot. 58: 639–646

Chitra DSV, Padmaja G. Clonal propagation of mulberry (Morusindica L. cultivar M-5) trough *in vitro* culture of nodal explants.Scientia Horticulturae. 1999;80(3): 289-298.

Dandin SB, Girish Naik V. Biotechnology in mulberry (Morus spp.) crop improvement: research directions and priorities. *Plant Biotechnology and Molecular Markers.* 2004; pp. 206-216.

Dandin SB, Jayaswal J, Giridhar K (2003) Mulberry cultivation. In: Dandin SB, Jayaswal J, Giridhar K (eds) Handbook of sericulture technologies. Central Silk Board, Bangalore, pp 1–5.

Das BC, Krishnaswami S. Estimation of component of variationof leaf yield and its related traits in mulberry. *J. Sericult. Sci.* 1969;38(3): 242-248.

Datta R.K. Mulberry Cultivation and Utilization in India. FAOElectronic conference on mulberry for animal production (MorusL). 2000.

Ford-Lloyd BV. The conservation of horticultural plants geneticresources. Plenary Lectures. 1990;87(4): 31-38.

Gamborg OL, Miller RA, Ojima K (1968) Nutrient requirements of suspension cultures of soybean root cells. Experimental Cell Research 50, 151-158.

Gamborg OL, Miller RA, Ojima K. Nutrient requirements ofsuspension cultures of soybean root cells. Experimental Cell Research.1968; 50: 151-158.

Guha A, Rasineni GK, Reddy AR (2010) Drought tolerance in mulberry: a physiological approach with insights to growth dynamics and leaf yield production. Exp Agric. 46: 471–488

H. T., Kester D. E., Davies F. T., Geneve Jr. R. L. (1997). Plant Propagation: Principles and Practices. 6th edition. Prentice-Hall, Inc., Upper Saddle River, N. J., 770 pp.

Kapur A, Bhatnagar S, Khurana P (2001) Efficient regeneration from mature leaf explants of Indian mulberry via organogenesis. Sericologia 41, 207-214.

Kartha KK. Production and indexing of disease-free plants, pp219-238. In Plant tissue culture and its agricultural applications (Eds., Lyndsey A. Withers and P.G. Alderson). 1986.

Kim HR, Patel KR, Thorpe TA (1985) Regeneration of mulberry plantlets through tissue culture. Botany Gazette 146, 335-340.

Kim HR, Patel KR, Thorpe TA. Regeneration of mulberry plantletsthrough tissue culture. Bot gaz. 1985; 146: 335-340.

Lloyd G, Mc Cown B. Commercially feasible micropropagation ofmountain laurel Kalmia latifolia by use of shoot tip culture.International Plant Propagation Society Proceedings. 1980; 30: 421-427.

Lloyd G, McCown B (1980) Commercially feasible micropropagation of mountain laurel Kalmia latifolia by use of shoot tip culture. International Plant Propagators Society Proceedings 30, 421-427.

Mandal BB. Conservation biotechnology of endemic and othereconomically important plant species in India. Plant conservationBiotechnology. 1999;70(5): 211- 225.

Mondal TK, Bhattacharya A, Laxmikumaran M, Ahuja PS (2004) Recent advances of tea (Camellia sinensis) biotechnology. Plant Cell, Tissue and Organ Culture 76, 195-254.

Murashige T, Skoog F (1962) A revised medium for rapid growth and bioassays with tobacco tissue cultures. Physiologia Plantarum 15, 473-497.

Murashige T, Skoog F. A revised medium for rapid growth andbioassays with tobacco tissue cultures. Physiologia Plantarum. 1962;15: 473-497.29. Von Arnold, S; Eriksson, T. *In vitro* studies on adventitious shootformation in Pinus contorta. Canadian Journal of Botany. 1981; 59: 870-874.

Niino T, Koyaman A, Shirata K, Ohuchi S, Suguli M, Sakai A. Longterm storage of mulberry winter buds by cryopreservation. Journal ofSericultural Science Japan. 1993;62(5): 431- 434.

Niino T, Sakai A, Enomoto S, Magoshi J, Kato S. Cryopreservation of*in vitro* grown shoot tips of mulberry by vitrification. Cryo-Letters.1992;13(2): 303-312.

Niino T, Sakai A, Yakuwa H. Cryopreservation of dried shoot tips ofmulberry winter buds and subsequent plant regeneration. Cryo-Letters. 1992;13(5): 51-58.

Niino T, Sakai A. Cryopreservation of alginate coated *in vitro*grown shoot tips of apple, pear and mulberry. Plant Science.1992;87(2): 199-206.

Niino T. Cryopreservation of germplasm of mulberry (Morus spp.). In: Y.P.S.Bajaj (ed.) Biotechnology in Agriculture and Forestry. Springer-Verlag Berlin. 1995;32(6): 102.

Ohyama K (1970) Tissue culture in mulberry tree. Japanese Agriculture Research Quarterly 5, 30-34.

Oka S, Ohyama K (1982) Sugar utilization in mulberry (Morus alba L.) bud culture. Proceedings of the 5th International Congress on Plant Tissue and Cell Culture, pp 67-68.

Ozkan Y., Arslan A. (1996). Karadin (*Morus nigra* L.) odunveyeþil çelikleçoðaltýlma sýüzerinearaþtýrmalar. Gazi Osman PaþaUniversitesi, ZiraatFakultesiDergisi, 13 (1): 15-27 (in Turkish).

Rao AA, Chaudhury R, Kumar S, Velu D, Saraswat RP, Kamble CK.Cryopreservation of Mulberry Germplasm Core Collection andAssessment of Genetic Stability through ISSR Markers. InternationalJournal of Industrial Entomology. 2007;15(1): 23 – 33.

Thomas TD. Thidiazuron induced multiple shoot induction and plantregeneration from cotyledonary explants of mulberry. BiologiaPlantarum. 2003; 46: 529-533.

Tingzing Z., Yunfang T., Guangxian H., Huaizhong F., Ben M. (1988). Mulberry Cultivation. FAO, Agricultural Service Bulletin 73/1, Rome, 127 pp.

Vijayan K, Ravikumar G, Tikader A. Advances in Plant BreedingStrategies: Fruits. In: AlKHayri JM, Jain M, Johnson DV. Mulberry (Morus spp.) Breeding for Higher Fruit Production. Springer.Bangalore. India. 2018;20(2): 89-130.

Vijayan K, Tikader A, da Silva AJT. Application of tissue culturetechniques for propagation and crop improvement in mulberry (Morus spp.). Tree Forest Science and. Biotechnology. 2015;1-13.

Vijayan K. Genetic relationships of Japanese and Indian mulberry (Morus spp.) revealed by DNA fingerprinting. Plant Sys Evol.2004;243(3): 221-232.

Von Arnold S, Eriksson T (1981) *In vitro* studies on adventitious shoot formation in Pinus contorta. Canadian Journal of Botany 59, 870-874.

Zaman A, Islam R, Joarder OI (1997) Field performance and biochemical evaluation of micropropagated mulberry plants. Plant Cell, Tissue and Organ Culture 51, 61-64.

Chapter 6

Conservation Strategies for Mulberry (*Morus* sp.) Genetic Resources

Shahina A.Nagoo*, Z.I. Buhroo, Sabiha Ashraf, S.A. Wani, and Marat Bashir

**College of Temperate Sericulture-Mirgund, Sher-e-Kashmir University of Agricultural Sciences and Technology of Kashmir, Srinagar – 190 025, J&K*
**Corresponding author e-mail: shahinanagoo.sn@gmail.com*

Diversity characterizes most of the life forms. Biodiversity covers variety of flora and fauna, their ecological roles and the genetic variability they contain. Exploitation of wild relatives of crop plants to a large extent depends on the efficient use of germplasm resources available in natural habitat and the center of diversity. The mulberry, sole food for silkworm is cultivated for the production and development of silkworm industry. Four species of mulberry viz., M. indica L., M. alba L., M. laevigata Wall., and M. serrata Roxb., are reported in India. Mulberry (Morus sp.) is one of them and is cultivated not only for sericulture but also for fruits, fodder, timber, fuel wood etc. Because of its adaptability to cross pollination with no inter-specific reproductive barriers a vast range of diversity in the genetic stocks has crept in thereby rendering it highly heterozygous plant species as a result every plant being different from other in natural population. To augment the diversity in true sense the characterization of the taxa is essential and the magnitude of diversity that has crept in the genetic stocks of the indigenous origin as regards to various qualitative characters, morphology, anatomy, propagation, reproduction, growth and biochemistry elucidated clearly signifies the wide range of diverseness in the indigenous mulberry genetic stocks. Conservation strategy unlike other field crops is yet to be refined for mulberry. However, means for conservation of diversity with probable risks in mulberry are well defined and have been discussed here under this book chapter.

Keywords: *Conservation, Mulberry (Morus sp.), Genetic resources.*

Introduction

The concept of biodiversity conservation and gene bank maintenance have gained great momentum in the recent times and the biodiversity wealth are considered as common heritage of mankind and "Sovereign Rights of the Nations". The issues related to access the plant genetic resources and its sustainable use, benefit sharing; farmer rights are being deliberated at various national and international fora. Plant Genetic Resources (PGR) comprises the total genetic diversity of the crop plants and their gene pool including adopted local landraces, wild relatives to elite breeders material, which forms a key component for sustainable development of any agricultural crop. India is recognised as one of the twelve-mega biodiversity centres of the world and harbouring 11.9 per cent of the world flora. Mulberry belongs to the fig family *Moraceae* under the order *Urticales* that is considered to be advanced in the woody flowering plants. Mulberry (Genus *Morus* L.) with high biomass production and protein rich foliage not only nurturing the entire sericulture industry in India besides, used extensively in agroforestry and horticulture programmes. India has attained the status of second largest silk producer in the world with galloping strides of raw silk production (15,000 metric tonnes accounting 16.1 per cent) after China. Mulberry leaf production has been tremendously increased from 15 to 60 metric tonnes/ha/year due to the incessant development of ideal plant types through intensive breeding system utilising diverse germplasm material. The need for diversity is continuous because of fast changing of the global environment. The scarcity of land and water will be the major constraints in future and development of varieties suitable for different agro climatic conditions are utmost important for sustainable development.

Realizing the importance of impending global climatic changes and threatened sustainability of biodiversity wealth in India at faster rate, systematic survey and exploration for collection of mulberry biodiversity have gained greater momentum in the recent past. Consequently, CSGRC at Hosur has been aptly established in the year 1990 by Central Silk Board (CSB), Ministry of Textiles, and Government of India under prestigious National Sericulture Project. It is the nodal agency for mulberry germplasm management in India and recognized as National Active Germplasm Site (NAGS) for mulberry by National Bureau of Plant Genetic Resources (NBPGR), New Delhi, India under Indian National Plant Genetic resources System (IN PGRS). CSGRC, Hosur so far conducted 46 survey and exploration trips covering more than 50 districts from Himalayan belt to Andaman and Nicobar Islands covering forest areas, biosphere reserves, national parks, back yards, kitchen gardens, agricultural lands and farmhouses *etc.* and collected 516 diverse mulberry germplasm resources which mainly belong to 4 Indian species (Table 6.1). Out of 4 species of genus *Morus* reported in India, *M.laevigata* is distributed throughout India both under natural and managed habitats and *M. serrata* is confined to northwest India in wild condition. *M. indica* and *M. alba* are mostly available in cultivated forms (Ravindran *et al.*, 1977). The sericulture research institutes like Central Sericultural Research and Training Institute (CSRTI) at Berhampore (West Bengal), Mysore (Karnataka) and Pampore (Jammu and Kashmir) have started collection of *Morus* species in Central Himalayas, North-Eastern India, North-Western Himalayas, Kumaon and North Eastern India and Western ghats and Kerala. All these germplasm resources are collected and introduced in the field gene bank.

Table 6.1: Current Status of Mulberry (*Morus* species) Germplasm Conservation in the World

Sl.No.	*Country*	*Total Species Available*	*Endemic Species*	*Total Germplasm Holdings*
1	China	24	17	2600
2	Japan	19	14	1312
3	South Korea	6	1	300
4	India	4	3	980
5	Indonesia	3	2	NA
6	Taiwan	4	1	NA
7	Thailand	2	2	NA
8	Argentina	1	1	NA
9	Columbia	3	1	NA
10	Mexico	3	2	NA
11	Peru	1	1	NA
12	USA	14	9	NA
13	France	NA	NA	70
14	Italy	NA	NA	50
15	Bulgaria	6	1	140

NA: Data is not available.

Source: Sanjappa, 1989.

Mulberry Genetic Resources at Global Level

Mulberry is native to temperate Asia and North America. Presently *Morus* L. grows in warm climatic zones between 50°N latitude and 10°S latitude (Yokoyama, 1962) which includes Sino-Japanese region of the old world and rocky mountains southwards to Andes of continental America in the new world and distributed over 29 countries or more. So far there are about 150 species of genus *Morus* have been described (Vide *Index Kewensis*, London) and presently 50 per cent of them have been reduced as synonymous or varieties of the same species. Today, *Morus* comprises about 68 recognised species and distributed in different countries mainly from Asian countries like Japan, China, India, Korea and Taiwan. Continental America is also rich in *Morus* species. The world distribution of genus *Morus* is given in Tables 6.1 and 6.2 (Sanjappa, 1989 and Dandin *etal*. 1999). Exploration of wild *Morus* species in crop improvement programmes has been greatly thought and

Table 6.2: Progressive Collection of Mulberry Germplasm through Survey and Exploration by CSGRC, Hosur

Explored Area Zone	*Indian State*	*Year-wise Mulberry Collection*			*Total Accessions Collected*
		I Phase Apr'93-Mar'97	*II Phase Apr'97-Mar'2000*	*III Phase April 2000 Onwards*	
North-West India	Jammu and Kashmir	-	-	12	12
	Himachal Pradesh	02	10	7	19
	Uttar Pradesh	66	40	-	01
	Uttaranchal	-	-	12	117
	New Delhi	-	02	-	02
	Punjab	-	01	7	08
	Haryana	-	04	-	04
North-East India	Assam	05	-	-	05
	Arunachal Pradesh	06	-	10	16
	Manipur	04	-	-	04
	Meghalaya	30	-	06	36
	Nagaland	01	-	-	01
	Sikkim	10	-	04	14
	West	14	-	07	21
	Bengal Jharkhand	-	-	18	18
Central India	Madhya Pradesh	14	-	-	14
	Orissa	01	-	-	01
Western India	Maharashtra	-	03	02	05
	Rajasthan	-	-	31	31
	Gujarat	-	-	30	30
South India	A and N Islands	02	44	-	46
	Andhra Pradesh	-	03	-	03
	Karnataka	-	14	-	14
	Kerala Tamil	30	10	-	40
	Nadu	30	24	-	54
Total		**215**	**155**	**146**	**516**

Source: Annual Report CSGRC, Hosur 2002.

major exploration activities in the Centres of Diversity were undertaken to collect wild species to supplement the genetic variability. Explicitly, Japan made spectacular advances in crop improvement utilizing putative genes form wild relatives of *Morus* species. *Morus multicaulis* species is widely used in many mulberry crop improvement programmes in Japan, India and china. The potentiality of American species and the isolated *Morus* species from different countries like Africa, Middle East, Thailand and Indonesia for alien gene transfer should also be looked into for exploitation.

Mulberry Genetic Resources: Indian Scenario

India exhibits wide variations in the altitude from sea level to the heights above vegetational limits in the Himalayas and extremely arid to semi arid in the north-western plains. It is floristically very rich with about 15,000 botanical wealth and out of this about 4 900 (33 per cent) are endemic, which belong to 141 genera, and 47 families distributed over 15 agroclimatic zones (Nayar, 1980 and Sehgal *et al.*, 1990). Great seri-biodiversity exists in the entire Sub Himalayan belt, north-eastern region, Central Plateau and Andaman and Nicobar Islands, which includes all the sericigenous insects and its host plants. Indian gene centre is very rich in *Morus* species. Wild relatives of genus *Morus* are reported to occur in India in tropical and sub tropical Himalayan belt from Indus to Arunachal Pradesh ascending to 1250 m. Rich *Morus* biodiversity also exists both under natural and managed habitats. Portuguese, British, French, Dutch and Mohammedan rulers in the process of colonization introduced many exotic varieties in the Indian gene centre. Besides, the introduced mulberry germplasm resources from different temperate and tropical countries, which are available in the Research institutes, Universities, and State departments, are also collected to enrich the mulberry field gene bank (Dandin *et al.*, 1992). This unique heritage serving human kind since ages has to be protected and conserved for the present and posterity.

In India, efforts have been made to use primary gene pools constituted predominantly cultivars belong to *Morus indica and Morus alba* to tailor the desired varieties. However, the promising genes exist in the secondary and tertiary gene pool in the wild species of *Morus laevigata* and *Morus serrata* and other wild species can be utilized for gene transfer by applying biotechnological and molecular techniques. Realizing the importance of mulberry germplasm for sustainable development of sericulture in India, Central Sericulture Germplasm Resource Centre (CSGRC) was established in India which is presently holding 908 diverse mulberry germplasm accessions representing 26 countries belong to 13 mulberry species in the *ex situ* field gene bank. Characterization and evaluation of 628 mulberry accessions has been completed and the trait specific promising accessions have been identified (Table 6.2).

Conservation of Mulberry Genetic Resources in India

Exploration for collection of mulberry biodiversity and development of efficient conservation strategies has gained momentum in the recent past in India. The conservation of mulberry genetic resources includes their augmentation, safe holding for medium and long-term preservation; protection in natural habitats and it is interlinked to the sustainable utilization of germplasm to justify long-term investments on managements and maintenance of the germplasm. Studying the geographical spread, distribution and genetic architecture, arborescent nature and physiological storage behaviour of mulberry, two basic conservation strategies *i.e. Ex situ* and *In situ* methods composed of various techniques covering entire gamut of genetic diversity have been worked.

Ex situ Conservation

Field Gene Bank

Mulberry being perennial out breeding tree exhibits high degree of heterozygosity. Hence, for conservation of mulberry outside its natural habitat, *ex situ* field gene bank (clonal repository) has been developed by planting vegetative clones of mulberry accessions for maintaining the genetic integrity of the conserved material. Rooted cuttings (saplings) after six-month establishment in the nursery are transplanted in the field gene bank. Saplings of some exotic accessions and wild *Morus* species with poor rooting rate, are developed through bud grafting to the local scions and then established in the base collection. The plants (four plants/ accessions) are maintained as a dwarf tree with spacing of 2.4 x 2.4 m between plants with the crown height of 1.5 m following recommended cultural practices and plant protection measures. Pruning is being followed once in a year (June-July) to renew the germplasm with new sprouts. However, the species, which cannot sustain repeated pruning are left unpruned. The entire collections in field gene bank are fenced and protected. Presently, the *ex situ* field gene bank of CSGRC, Hosur holds 908 mulberry accessions (Indigenous-647 and Exotic-261) from diverse genetic and geographical origin representing 13 *Morus* species collected from 26 countries. CSGRC, Hosur being the nodal agency for mulberry germplasm in India maintain the entire mulberry germplasm available in the country. However, the Sericulture Research Institutes, Universities, State Sericulture Departments are also maintain the mulberry germplasm for research purpose, which also serves as backup conservation centres.

On-farm Participatory Conservation

The seri-biodiversity otherwise greatly threatened because of unlawful habitat destruction, natural calamities, fragmentation of forests and social disruption and this large scale genetic wipeout disturb the coexistence of sericigenous flora and fauna. Under these circumstances on-farm conservation linked with Farmers Participatory Breeding (FPB) to be given due emphasis for achieving twin goals of sustainable conservation of biodiversity and in turn utilization on-farm biodiversity (Eyzaguirre and Iwanaga, 1996). In India, rich *Morus* diversity exists under managed habitats *i.e.* in the backyards, kitchen gardens, farmhouses, horticultural gardens, agricultural lands and roadside plantations. These are the firsthand selections of the farmers and tribal's for varied utilizations hence; conservation of potentially interesting alleles and development of diversity is promoted. In mulberry the wild species like *Morus laevigata* and *Morus serrata* and other wild species, which do not get attention in the formal sector for cultivation for sericulture purposes.

So, these associated valuable species or otherwise utilized for non-sericultural purposes (horticulture and agroforestry) flourish well in the on-farm conservation procedures promoting farmers/tribal's livelihood development while conserving *Morus* biodiversity. This particular sector of *Morus* biodiversity lies mainly in the public domain, which needs to be taken care and bring them under definite set of legal framework of Indian Forest Act (1972).

Botanical Gardens and National Herbarium

Seri-biodiversity museum of CSGRC, Hosur maintains large number of herbaria and it serves as National Herbarium Centre for *Morus* species in India. In India there are about 33 botanical gardens and the some of them like: Indian Botanical garden, Howrah; Lloyd Botanical garden, Darjeeling; National Botanical Garden, Lucknow maintain mulberry arboreta.

Conservation Using Biotechnology *In vitro* and Cryopreservation

Cryopreservation technology forms an alternative to *ex situ* conservation strategy for crop like mulberry, which is vegetatively propagated utilizing liquid nitrogen (LN) at ultra low temperature of -150 °C (vapour phase). Cryopreservation method ensures the genetic stability of the germplasm and provides an alternate cost effective non-dependence on electricity long-term strategy for conservation of mulberry germplasm. Keeping this in view, establishment of *in vitro* and cryopreservation laboratory facilities at CSGRC, Hosur is planed in collaboration with NBPGR, New Delhi. Efforts have been made to standardize the efficient techniques on cryopreservation of winter buds, embryonic axes, pollen of mulberry accessions (Niino *et al.* 1992 a, b, c, 1993 and 1995) comprising different species, landraces, local cultivars, wild and polyploid accessions in liquid nitrogen (-196°C). *In vitro* conservation techniques involve conservation of active collections under active growth stage and base collection germplasm under suspended growth stage using growth retardants.

Cryopreservation procedures have been standardised for about 100 different plant species cultured in various ways including cell suspension, callus, apices and zygotic and somatic embryos. Cryopreservation, possibility for the first time was demonstrated using mulberry twig by Sakai (1960). Since then, considerable work on cryopreservation of mulberry has been undertaken especially in Japan. In India, the application of *in vitro* technique for mulberry conservation has been recently attempted. *In vitro* technique has been attempted in mulberry mainly for propagation of popular mulberry varieties and poor rooting materials, evolution of some clonal variants and elite materials and screening of genotypes for tolerance to salt and somatic stresses. (Tewary *et al.,* 1996 and 2000). Bapat *et al.* (1987) developed method for propagation of *Morus indica* L. (Mulberry) by culturing encapsulated shoot buds. Conservation of mulberry germplasm accessions through cryopreservation technique has been recently attempted in India.

In situ Conservation of Mulberry Genetic Resources

In situ conservation demands the establishment of nature or biosphere reserve and national parks to protect the endangered species. Keeping this in view, efforts have been made to collect information on the location of availability of mulberry germplasm with details on "declared protected area network of India" including biosphere reserves, national parks, wild life sanctuaries *etc.* Mulberry is not fully protected under Indian Forest Act (1972) in many parts of India except in some states of North East India. Survey map, exact location, landowner with postal address, survey number and its jurisdiction are not available with CSB units or State department. Under the circumstances, a suggestive and advisory role can be contemplated with greater thrust on repeated survey and exploration.

Since ages, mulberry is being worshiped by tribals of Uttaranchal of Himalayan region and conserved in the name of "Sacred grooves". The sacred mulberry tree (*M. serrata*) at Joshimath (altitude of 2 000 above MSL) being worshipped by the pilgrims of Badrinath, is said to be the oldest (1200 years) and biggest mulberry tree (21.6 m of circumference) in the world. Attempts are being made to locate the concentrated areas of wild/natural mulberry trees of *Morus laevigata* at Island of Andaman and Nicobar biosphere, Arunachal Pradesh and areas of Himalayan region. Large numbers of naturally grown *M.laevigata* trees are also available in Eastern Himalayas of south and west Sikkim, West Bengal, Assam, Meghalaya and Manipur. Besides, developed gene pool of *M.alba, M.indica* and *M.laevigata* are also available in western ghats covering Kerala, Tamilnadu and Karnataka, Madhya Pradesh, Maharastra, Rajasthan. Yarcaud hills of Tamilnadu and Simlipal reserve forest of Orissa *etc.* which are adapted to local environment for many years are also required to be protected under *in situ* conservation.

Genetic Diversity and Characterization of Mulberry

Molecular Characterization

Molecular markers on the other hand exhibit high degree of non-tissue specific polymorphism and simple inheritance pattern with minimum influence of environmental and epistasis. In the last two decades, geneticist has started using biochemical and molecular techniques to add valuable information for characterization of germplasm. In addition to the fact this genetic information can be generated quickly, they are environmentally independent, accurate and reliable. Isozyme markers have been successfully used to characterise plant genetic resources. Even though, isozyme analysis is simple and relatively cost effective they are confine to small portion of genome and often they do not offer sufficient polymorphism to reveal the genetic difference. In the pursuit, molecular approaches like RFLP, RAPD, microsatellite markers, AFLP *etc.* effectively overcome this problem by offering high level of polymorphism. These techniques along with computer based software programmes are being routinely used in characterization and germplasm related studies like marker based gene tagging, gene cloning, diversity analysis and marker assisted selection of desirable genotypes all over the world.

Isozyme Analyses

At CSGRC, Hosur the isozyme study was initiated in 17 representative *Morus* species using peroxidase enzyme system to assess the allelic variation at isozymic loci and to ascertain the genetic diversity. All the species showed one common band but variability was restricted to minor bands. Based on the number of bands and Rf values, these species were categorised into six groups. Fifty mulberry landraces and popular varieties were subjected to Isozyme analysis using three enzyme system namely anodal peroxidase, acid phosphotase and cathodal peroxidase to study the extent of genetic diversity. Peroxidase banding pattern showed greater variability. *M. laevigata* is distributed in different parts of the country both in wild and managed habitats. In order to assess the genetic diversity in 32 *M. laevigata* explored collection were subjected to isozyme analyses. Peroxidase, β- esterase and Cathodal peroxidase enzyme systems showed wider genetic polymorphism. Acid phosphotase enzyme again did not show any genetic polymorphism among 32 *M.laevigata* accessions. Isozyme banding pattern did not show any correlation with geographical distribution of collections. However, *M.laevigata* collections from north-eastern India revealed maximum genetic diversity at isozymic loci indicating potentiality for further exploration and collection. Polymorphism at Isozyme loci is comparatively low in mulberry as is the case in most of the cultivated crops. Hence, there is a need to take up more number of enzyme systems for Isozyme analyses to obtain conclusive result (Anonymous, 1997-1998).

DNA Fingerprinting

Molecular characterization of mulberry germplasm was initiated in collaboration with Seri Biotech Research Laboratory (SBRL), Bangalore. Initially, protocol for mulberry DNA extraction, RAPD and Inter SSR- PCR analysis was standardised. Fifteen morphological distinct species were analysed using RAPD and Inter SSR-PCR assays. In RAPD assay, 19 oligonucleotide primers were utilised. These analyses also identified few species-specific amplification products. Inter SSR-PCR analyses were carried out in 15 mulberry species using four primers. However, the species-specific markers have to be studied over populations of individual species so as identify the unknown species. The RAPD analyses of 46 mulberry varieties using five oligonucleotide primers revealed polymorphism suggesting the genetic variation. However, clustering pattern did not show correlation with regard to regional/geographical distribution of these varieties (Chatterjee and Awasthi, 2000). National Botanical Research Institute, Lucknow studied nine mulberry germplasm varieties (triploids, species and popular varieties) using 23 arbitrary sequence decamer primers for RAPD and three minisatellite core sequence primers for DAMD polymorphism. Nearly 85 per cent RAPD and 91 per cent DAMD bands were polymorphic across nine varieties differentiating ployploids and popular varieties (Bhattacharya and Ranade, 2001).

Future Directions in Conservation of Mulberry Genetic Resources

The following recommendations are suggested for management of mulberry germplasm and utilization at global level.

- ☆ Establishment of International Networking on Sericultural Germplasm Management at global level, Co-ordination between countries with continuous interaction and linkages and identification of working group with international training and technology transfer under auspices of FAO.
- ☆ Documentation of total mulberry germplasm holdings with detailed passport information at global level to develop international database on mulberry germplasm.
- ☆ Standardization of formats for passport information, characterization and evaluation measurement methods and publication of mulberry descriptor in English
- ☆ Report generation on status of mulberry germplasm characterization and evaluation and gap analysis and development of mulberry core collections.
- ☆ Taxonomical revision of genus *Morus* at global level to be undertaken. There is an urgent need for congregation of all *Morus* species available in the world at one place, which will facilitate to revise taxonomical position of genus *Morus* at global level. Monographing of genus *Morus* is highly essential.
- ☆ Joint survey and exploration for collection of wild mulberry germplasm resources, mapping and indexing of seri-biodiversity.
- ☆ Finalizing the Global Action Plan for Seri-biodiversity.
- ☆ International screening of mulberry germplasm and identification of promising lines for exploitation.
- ☆ Inclusion of mulberry in the Multilateral System for Global Exchange with benefit sharing. In the beginning the wild genetic resources and different *Morus* species native and endemic to respective countries, which are hitherto, are not developed may be shared to conduct studies on wide hybridisation and utilization.
- ☆ International cooperation for funding mechanism/monitoring by FAO, UNDP, WWF and other organization for conservation and management of mulberry genetic resources.
- ☆ Establishment of *In vitro* and Cryopreservation facilities for long-term conservation of mulberry germplasm, which will also facilitate global exchange of mulberry germplasm. International training at Japan is suggested as it has done excellent work on Cryopreservation of mulberry germplasm resources.

- ☆ Promotion of *in situ* conservation with *ex situ* backup to be given high priority
- ☆ International networking of sericultural website and linkage with international genetic resources websites like WIEWS, SINGER, UKPGRG, NGB, GRIN and IN ARIS for easy accesses of mulberry germplasm information and utilization of the resources.

REFERENCES

Anonymous, 1975. *Text book of Sericulture*, Japan Overseas Co-operation Volunteers, 4-2-24 Hiroo, Dibuys-Ku, Tokyo, Japan.

Anonymous, 1997-1998. *In*: *Annual Report*, Central Sericultural Germplasm Resources Centre, Central Silk Board, Hosur, Tamil Nadu. pp.36-37.

Bapat, V.A., Mhatra. M and Rao, P.S. 1987. Propagation of *Morus indica* L. (Mulberry) by encapsulated shoot buds. *Plant Cell Reports.* 6: 393-395.

Bhattacharya, E. and Ranade, S.A. 2000. Molecular distinction amongst varieties of mulberry using RAPD and DAMD profiles. *BMC Plant Biology*. 1: 3. (available at www.biomedicentral.com).

Bindroo, B.B., Fotadar, R.K. and Dhar, A. 1996. Propagation of mulberry under subtropical conditions. *Indian Silk.* 34 (9): 11-15.

Chatterjee, S.N. and Awasthi, A.K 2000. Molecular characterization of germplasm- An over review. *In*: *National workshop on management of sericultural germplasm for posterity* (*Base papers*), 26th -27th, July, 2000, CSGRC, Central Silk Board, Hosur. pp: 39-49.

Chinnaswamy, K.P. and. Hariprasad, K.B 1995. Fuel energy potential of mulberry. *Indian Silk.* 34(4): 15-18

Dandin, S.B. 1997. Mulberry improvement programmes- An overview. *Indian Silk*, pp.5-9.

Dandin, S.B., Mala V R. and Mallikarjunappa, S. 1992. Conservation of Genetic resources of mulberry. *In*: R. Saxena, G. Rammurti, SreeramaReddi and V.L Saxena. Edi. *Recent advances in life sciences,* Indian Society of Life Sciences, Manu publication, Kanpur, UP, India. pp. 223-236.

Dandin, S.B.1999. Mulberry improvement programme for sustained productivity under tropics.In: M.C Devaih, K.C.Narayanaswamy and V.G.Mariba Shetty edi. *Advances in mulberry sericulture,* C.V.G. Publication, Bangalore, India. pp. 51-87.

Dorcus, D. and Vivekanandan, M. 1997. Exploitation of mulberry genotypes for drought resistance potential. *J. Seric. Sci.,* 66 (2): 71-80.

Goel A.K., Ravindran, S, Ananda Rao, A, Naik, V.G, Tikader, A., Mukherjee, P. and Sekar, S. 1998. Variability in rooting parameters at juvinile stage in mulberry (*Morus* spp.) germplasm. *Indian J. Seric.,* 37(2): 109-112.

Katsumata, T. 1972. Mulberry species in West - Jawa and their peculiarities. *J. Seric. Sci.*, Japan, 42: 213-223.

Koidzumi, G. 1917. Taxonomy and Phytogeography of the genus *Morus*. *Bull. Seric. Expt. Stn.*, Tokyo. 3 (1): 1-62.

Ledebour, C.F. 1951. Flora Rossica Siveenumerhtio Plantarum in TotiusImperii Rossici Symptilus Librarial E. Schuscizerbart. *Stuttgartiae*, 3: 643-644.

Linnaeus, 1753. *Species Plantarum*: pp 986

Nair, N.C. 1977. *Flora of Bushahr Himalayas*, International Bioscience Publishers, Hisar.

Nayar, M.P. 1980. Endemism and patterns of distribution of endemic genera (angiosperms). *J. Econ. Tax. Bot.*, 1: 99-110.

Niino, T and Sakai, A. 1992 a. Cryopreservation of alginate - coated *in vitro* grown shoot tips of apple, pear and mulberry. *Plant Sci.*, 87: 199-206.

Niino, T. 1995. Cryopreservation of germplasm of mulberry (*Morus sps.*). *In*: Y.P.S.Bajaj (ed.).*Biotechnology in Agriculture and Forestry*, Springer-Verlag, Berlin.32: 102-113.

Niino, T., Koyaman, A, Shirata, K., Ohuchi, S., Suguli, M. and Sakai, A. 1993. Long term storage of mulberry winter buds by cryopreservation. *J. Seric. Sci.* Japan. 62(5): 431-434.

Niino, T., Sakai, A. and Yakuwa, H. 1992 b. Cryopreservation of dried shoot tips of mulberry winter buds and subsequent plant regeneration. *Cryo-Lett.*, 13: 51-58.

Niino, T. Sakai, A., Enomoto, S, Magoshi, J&Kato, S. 1992 c. Cryopreservation of *in vitro* grown shoot tips of mulberry by vitrification. *Cryo.Lett.*, 13: 303-312.

Ramesh, S.R. and Basavaiah 1989. Systematic position of the genus *Morus* L. In: K.Sengupta and S.B.Dandin (Eds.), *Genetic Resources of Mulberry And Utilization*, CSRTI, Mysore.pp.43-48.

Rao Ananda, A., Tikader, A, Ravindran, S. and Mukherjee, P. 2000. Sexual diversity in mulberry (*Morus* Spp.) germplasm and its utilization. *Indian Jour. of Forestry.* 23(3): 312-318.

Rao Ananda, A., Tikader, A. and Thangavelu, K. 2002. Variation in mulberry germplasm and their utilization. *Indian Silk*. 41(4): 5-9.

Ravindran, S., Ananda Rao, A., Naik, V.G., Tikader, A., Mukherjee, P and Thangavelu, K 1997 Distribution and variation in mulberry germplasm, *Indian J. PL. Genet. Resources.* 10(2): 233-242.

Ravindran, S., Tikader, A., Naik, V.G., Ananda Rao, A. and Mukherjee, P. 1999. Distribution of mulberry species in India and its utilisation. *Indian J.PL Genet. Resources*, 12(2): 163-168.

Sakai A. 1960. Survival of the twigs of woody plants at -196°C. *Nature*. 185: 392-394.

Sanjappa, M. (1989). Geographical distribution and exploration of the genus *Morus* L. (*Moraceae*). In: K.Sengupta and S.B.Dandin (Eds.) *Genetic Resources of Mulberry And Utilization*, CSRTI, Mysore, pp: 4-7

Sekar, S., Goel, A.K., Mukherjee, P. and Thangavelu, K. 1997. Management of information system in germplasm maintenance. *Indian Silk*. October 6-9.

Tewary, P.K., Chakraborty, S. and Subba Rao, G. 1989. Taxonomy of the genus Morus L. - A critical appraisal. In: K.Sengupta and S.B.Dandin, eds. *Genetic Resources of Mulberry And Utilization*, CSRTI, Mysore, pp. 49-54

Thangavelu, K., Mukherjee, P., Tikader, A. Ravindran, S., Goel, A.K., Ananda Rao, A., Girish Naik, V. and Sekar, S. 1997. *Catalogue on mulberry* (*Morus spp.*) *germplasm* Vol.1, SMGS, CSB, India.pp.236.

Tikader, A. Ananda Rao, A., Ravindran, S., Girish Naik, V., Mukherjee, P. and Thangavelu, K. 1999. Divergence analysis in different mulberry species. *Indian J. Genet.* 59(1): 87-93

Tikader, A. and Ananda Rao, A. 2001. Analysis of divergence in mulberry species through different clustering technique. *Sericologia*, 41(1): 109-116.

Vavilov, N.I. 1926. Studies on the origin of cultivated plants, *Trudy ByuroPrikl. Bot.*, 16: 139-248.

Chapter 7

Diversity, Distribution and Abundance of Earthworms from Ashti Region Dist. Beed (MS)

V.V. Vaidya

Department of Zoology, Smt. S.K. Gandhi Arts, Amolak Science and P.H. Gandhi Commerce College, Kada, Tq. Ashti, Dist. Beed Maharashtra

Distribution pattern of earthworm fauna in Ashti region and its population densities were studied. The data on relative density, frequency and abundance of four types of earthworm species found at soil of Ashti Tahshil. Among these Polypheretima elongata, Pheretima posthuma, Lampito mauritii and Perionyx sansibaricus were found widely distributed with 36.97 per cent, 34.45 per cent, 18.48 per cent and 10.08 per cent relatively density and 45.45 per cent, 41.66 per cent, 24.39 per cent and 22.72 per cent Lampito mauritii, Perionyx sansibaricus, Pheretima posthuma and Polypheretima elongate relative abundance. It was observed that highest density was found in the species Polypheretima elongata and lowest distributed earthworm species was Perionyx sansibaricus as comparable to other two worms

Keywords: *Abundance, Distribution, Diversity, Earthworm, Ashti region.*

Introduction

Earthworms, the soil macroinvertebrates, are prominent among soil fauna and regulate the soil processes (Ismail 1997). They are found in all types of soils with sufficient moisture and food (Ghosh 1993). They act as decomposers and also a rich protein source (Neuhauser *et al.,* 1979). Belonging to class Oligochaeta, earthworms form the major terrestrial and soil inhabiting organisms of Phylum

Annelida (Ghosh 1993). In India, Julka (1993) reported 509 species and 67 genera of earthworms. Review of literature reveals that earthworm population dynamics in relation to different soil types is known (Dash and Patra 1977; Fragoso *et al.*, 1999). Ghosh (1993) presented the ecological aspects of Indian earthworms along with their vermicomposting technique. However, there is no report on distribution and abundance of earthworms in Ashti region. To fill in this lacuna, the present study was undertaken. The study focuses on population density, species wise age and depth distribution in different habitats of different localities.

The word "biodiversity" has exploded in Governmental and intergovernmental reports, news papers, meeting reports, and vocabulary of popular press during the last two decades. Growth in usage of this term has been dramatic in the scientific literature. There is long history of biological diversity in variety of contexts. Biodiversity can be considered synonymous with biological diversity for practical purpose as defined by Norse *et al.* (1986). This is reinforced by the official definition in the convention on "Biological Diversity" signed by 156 countries and European community at the United Nations Conference on the Environment and Development *i.e.* The Earth Summit in 1992 at Rio de Janeiro in Brazil. "Biological diversity" means the variable among living organism from all sources including terrestrial and aquatic systems and ecological complexes of which they are part; this includes diversity within species and of ecosystem.

The distribution and abundance of species are determined by a combination of factors including biogeographically history, environmental condition (*e.g.* temperature, salinity, storms), availability of spatial and food resources, rates of reproduction, mortality and migration and intra- and inter species interactions (*e.g.* predator and competitor) (Branch, 1976; Underwood, 1988). At local scale the availability and use of resources and interactions with other individuals are key factors regulating the morphology and behavior of individuals and these in turn affect population (Hopkins and Gartner 1992; De'ath and Moran, 1998).

The diversity index gives a measure of the individual present in a community and global loss of thousands of species as a result of population and habitat destruction. Such studies assist environmental biologists to predict where and how many species go extinct so that certain effective measures may be undertaken to protect them. Not much information is available on these aspects regarding the population of the earthworm from this region.

The present study area from the Ashti region is an attempt to define the status of the species and to establish some nomenclatural stability. It is also cautionary in that taxonomic progress can only be made by resolving, as far as possible, the inherited conflicts in species descriptions.

Materials and Methods

Application of statistical methods to biological data as biostatistics or biometry is an important tool in biological research. It generates useful information for interpretation of the results obtained during biological experimentation of the research and development. Population densities of earthworm species were estimated by quadrate method of Greig and Smith (1983). Quadrate is a unit area of square rectangular or circular quadrate of unique length and radius respectively. This can be used to study the sessile and fixed animals.

It is necessary to determine the "minimum size of quadrate", before proceeding to study the community, so it can be used in field application for community study. Presently in biodiversity measurement and community composition study, the quadrate method has its own importance and frequently used in identification of relative abundance of species and quantification of the resources. The present study was carried out at 10 selected sites from Ashti resgion soil during October 2013 to August 2014. To study earthworm species distribution of a series of 10 quadrates (1m X 1m) were laid at an interval of 200 m along soil of Ashti resgion. The relative density, relative frequency and relative abundance were computed following the method of Sutherland (1997).

1. **Relative density** $= \dfrac{\text{No. of individual species}}{\text{No. of individuals of all species}} \times 100$

2. **Relative Frequency**

$$= \frac{\text{Total no. of quadrates in which species occurred}}{\text{Total no. of quadrates studied}} \times 100$$

3. **Relative Abundance**

$$= \frac{\text{Total no. of quadrates in which species occurred}}{\text{Total no. of quadrates of species occurrence}} \times 100$$

Results

The data on relative density, frequency and abundance of four types of earthworm species found at soil of Ashti resgion are presented in Table. Among these *Polypheretima elongata, Pheretima posthuma, Lampito mauritii* and *Perionyx sansibaricus* were found widely distributed with 36.97 per cent, 34.45 per cent, 18.48 per cent and 10.08 per cent relatively density and 45.45 per cent, 41.66 per cent, 24.39 per cent and 22.72 per cent *Lampito mauritii, Perionyx sansibaricus, Pheretima posthuma* and *Polypheretima elongate* relative abundance. It was observed that higest density was found in the species *Polypheretima elongata* and lowest distributed earthworm species was *Perionyx sansibaricus* as comparable to other two worms.

Discussion

Soil fauna is influenced by environmental factors and changes in soil physical and chemical properties of soil influence their number and local distribution patterns (Ettema *et al.,* 1998, 2002). Some of these changes, such as seasonal variations, are time-related, affecting animal life cycles and their annual population dynamics. Moreover, differences in resource availability also play an important role to terminal some life history traits such as growth, size at maturity, number and size of offspring and length of life (Stearns, 1992). Earthworms represent a major group into the soil fauna and seasonal factors play an important role in explaining changes in size and biomass of their populations (Edwards and Bohhlen, 2004). Because earthworm species exhibit different ecological preferences, the influence of environmental factors on population dynamics differs among earthworms of different ecological categories (Rozen, 1998).

The soil conditions that affect earthworm activity like temperature, moisture and food (Edwards and Bohhlen, 2004). However epigamic earthworm species are found living and feeding in the litter layer. In the present study, the distribution of three species *viz. Polyperetima elongate, Lampito mauritii* and *Periyonyx sansiburcus* were recorded. The species richness observed in the present study is low in comparison to those reported in other parts in India such as seven species by Kale and Krishnamoorthy, (1981), six species by Ismail, (1986), five species by Mohanjit, (1986), and Tiwari (1983). Observation of the present study demonstrated that all species of earthworm occurring in Ashti were widely disperced with respect to C:N ratio which coincides with the observation of Kale and Krishnamoorthy (1981). The study revealed that the earthworms Polyperetima elongate and Lampito mauritii were abundantly distributed in the soil, where C:N ratio is higher and Perionyx sansiburcus was abundantly distributed in the soil rich in organic carbon and nitrogen contents leading to low C:N ratio.

The basic aspects of the life cycle of *Eisenia fetida* are relatively well documented due to the importance of this species in waste management. The growth and reproduction of this earthworm species is affected by several factors such as food quality, moisture, temperature and population density, as showed under laboratory conditions (Diminguez, 2004). Partly due to the particular distribution of this species, mainly associated to compost and manure piles (Bouche, 1972; Watanable and Tsukamoto, 1976).

In the present work, *Polypheretima elongata, Pheretima posthuma, Lampito mauritii* and *Perionyx sansibaricus* were found widely distributed with 36.97 per cent, 34.45 per cent, 18.48 per cent and 10.08 per cent relatively density and 22.72 per cent, 24.39 per cent, 45.45 per cent and 41.66 per cent relative abundance. Lowest distributed earthworm species was *Perionyx sansibaricus.* The *Polypheretima elongata* is the largest group of earthworms in the world, consisting of more than 700 nominal species and subspecies (Sims and Easton1972). It is also the largest group in India. Because it is too large to be handled, Sims and Easton (1972) divided the genus into 8 genera by phonetic analysis according to the greatest number of shared morphological attributes.

Lavelle (1978) reported the species *Millsonia lamtoiana* ranged in density from 0.02 to 1.43 ind.m^{-2} and biomass from 0.01 to 8.43g fresh weight mb^2 in the Ivory Coast, rather high values are not match with *Millsonia lamtoiana*. Soil moisture is the most important of all environmental variables for earthworms in tropical soils (Lavelle, 1983). Garnsey (1994) also reported earthworm biomass in the Midlands of Tasmania, *Millsonia carininguensis* has a patchy distribution pattern with high density spots alternating with low density areas. Earthworm aestivation has been observed throughout temperate regions (Nordström, 1975; Anderson, 1980) In Africa (Madge, 1969) and in Australia (Garnsey, 1994), grasslands are known to support high earthworm populations and biomass (Lee, 1985). Litter composition is the primary factor determining these high values (Mishra and Ramakrishnan, 1988). Differences in the litter quality, and large inputs of cattle dung which is rapidly incorporated into the soil by this species and dung beetles, may also be factors responsible for the enhancement of earthworm activity in improved pastures.

Understanding the earthworm diversity of any region requires consideration of the species recorded from adjacent countries as distributions often overlap borders. Gates (1972) described about 174 mainly pheretimoid species from Myanmar (Burma), with much information on their relationships to other Oriental faunas. In contrast, the British Island of similar to Japan have only 48 taxa comprising species that are unlikely to be entirely endemic, having colonized after the last ice age, and many of which are also recorded as common exotics elsewhere (Sims and Gerard 1985, 1999). Approximately 160 species are known from North America that was also widely calculated. 350 species are reported from India; 193 species (170 native) from NewZealand (Rossi, 2003).

REFERENCES

Andrade, C. Canitrot, E. Barreiros P, and Salsón S, (2000): Anuario Climatolóxico de Galicia, Consellería de Medio Ambiente, Xunta de Galicia, Santiago de Compostela.

Andrade, C., Canitrot, E., and de Galicia (1999), Consellería de Medio Ambiente, Xunta de Galicia, Santiago de Compostela,

Baker GH, Barrett V J, Grey-Gardner R, Buckerfield JC (1992) The life history and abundance of the introduced earthworms Aporrectodea trapezoides and A. caliginosa (Annelida: Lumbricidae) in pasture soils in the Mount Lofty Ranges, South Australia. Australian Journal of Ecology 17, 177-188.

Baker GH, Lee KE (1993) Earthworms. In 'Soil sampling and methods of analysis'. (Ed. MR Carter) pp. 359-378. (Lewis Publishers: Boca Raton, FL)

Barnes BT, Ellis FB (1979) The effects of different methods of cultivation and direct drilling and of contrasting methods of straw dispersal on populations of earthworms. Journal of Soil Science 30, 669-679.

Belmap, J., Kaltecer J., Rosentreper, R, Willimms J., Leonrd, S. and Eldridge D. (2001): Biological soil crusts: Ecology and management. WWW. Soil crust.

Blair GJ, Lefroy RDB, Lisle L (1995) Soil carbon fractions based on their degree of oxidation, and the development of a carbon management index for agricultural systems. Australian Journal of Agricultural Research 46, 1459-1466. doi: 10.1071/AR9951459

Bohlen PJ, Edwards WM, Edwards CA (1995) Earthworm community structure and diversity in experimental agricultural watersheds in Northeastern Ohio. The significance and regulation of soil biodiversity. Plant and Soil 170, 233-239. doi: 10.1007/BF02183060

Boone FR, Slager S, Miedema R, Elevld R (1976) Some influence of zero-tillage on the structure and stability of a fine-textured river levee soil. Netherlands Journal of Agricultural Science 24, 105-119.

Brady N. C. (1990): the nature and properties of soil. 10th Edn. Mcmillan publishing company New york

Chan KY (2001) An overview of some tillage impacts on earthworm population and diversity–implications for functioning in soils. Soil and Tillage Research 57, 179-191. doi: 10.1016/S0167-1987(00)00173-2

Cluzeau, D., Fayolle, L and Hubert M (1992): The adaptation value ofreprod0uctive strategy and mode in three epigeous earthworm species, Soil Biol. Biochem. 24 (1992) 1309–1315.

Dash M. C. and Senapati B. K. (1980): Cocoon morphology, hatching emergence pattern in tropical earthworm. Pedobilogia, 20: 316-324.

Dash M. C., Senapati B. K. and Behera N. (1979): Microfungi associated with decomposition of earthworm tissue in pasture soil. Biol. Bull. India: 1: 21- 23.

Dash M. C., Senapati B. K. and Misra C. C. (1980): Nematode feedind by tropoical earthworms. Oikos, 34: 322-325.

Dash, M. C. and Patra, U. C. (1977): Density, biomass and energy budget of tropical earthworm population from a grass land sight in Orissa, India. Rev. Ecol. Boil. Sol. 14; 461-471.

Dhaliwal G. S., Sangha G. S. and Ralhan P. K. (1996): Fundamentals of environmental sci. Kalyani publishers Ludhiyana.

Domínguez, J. (2004): State of the art and new perspectives on vermicomposting research, in: C.A. Edwards (Ed.), Earthworm Ecology, CRC Press, Boca Raton, FL, USA, 2004, pp. 401–424.

Doube BM, Buckerfield JC, Kirkegaard JA (1994) Short-term effects of tillage and stubble management on earthworm populations in cropping systems in southern New South Wales. Australian Journal of Agricultural Research 45, 1587-1600. doi: 10.1071/AR9941587

Easton EG (1979). A revision of the 'acaecate' earthworms of thePheretima group (Megascolecidae: Oligochaeta): *Archi*-pheretima, *Metapheretima,*

Planapheretima, Pleionogaster and *Polypheretima*. Bull. Br. Mus. Nat. Hist. (Zool.) **35:** 1-126.

Easton EG (1982): Australian pheretimoid earthworms (Megascolecidae: Oligochaeta): a synopsis with the description of a new genus and five new species. Aust. J.Zool. **30:** 711-735.

Edwards C. (1998): Earthworm ecology. Delray Beach Flaa: St. Lucie press.

Edwards CA, Bohlen PJ (1992) The effects of toxic chemicals on earthworms. Reviews of Environmental Contamination and Toxicology 125, 23-99.

Edwards CA, Bohlen PJ (1996) 'Biology and ecology of earthworms.' (Chapman and Hall: London)

Edwards CA, Lofty JR (1982) The effects of direct drilling and minimal cultivation on earthworm populations. Journal of Applied Ecology 19, 723-734.

Edwards CA, PJ Bohlen. 1996. Biology and ecology of earthworms. London: Chapman and Hall.Gates GE. 1937. Indian earthworms. I. The genus Pheretima. Rec. Ind. Mus. **39:** 175-212.

Edwards, C. A. and Bohlen P.J. (2004): Biology and Ecology of Earthworms, Chapman and Hall, London.

Ettema, C. H., Coleman, D C. Vellidis, G. Lowrance, R and Rathbun, S.L (1998): patiotemporal distributions of bacterivorous nematodes and soil resources in a restored riparian wetland, Ecology 79: 2721–2734.

Ettema, C.H. and Wardle, D. A. (2002): Spatial soil ecology, Trends Ecol. Evol. 17: 177–183.

Fragoso, C., P. Lavelle, E. Blanchart, B.K. Senapati, J.J. Jimene, M.A. Martinez, T.Decaens and J.Tondoh. 1999. Earthworm communities of tropical agroecosystems: origin, structure and influence of management practices. pp. 27-55. In: Earthworm Management in Tropical Agroecosystem. CAB International. Wallingford, U.K.

Francis GS, Knight TL (1993) Long-term effects of conventional and no-tillage on selected soil properties and crop yields in Canterbury, New Zealand. Soil and Tillage Research 26, 193-210. doi: 10.1016/0167-1987(93)90044-P

Gates G. E. (1972): Burmese earthworms. An introduction to the systematics and biology of megadrile oligochaetes with special reference to Southeast Asia. Trans. Am. Phil. Soc., N.S. **62:** 1-326.

Gerard BM, Hay RKM (1979) The effect on earthworms of ploughing, tined cultivation, direct drilling and nitrogen in a barley monoculture system. Journal of Agricultural Science, Cambridge 93, 147-155.

Ghosh, A.H. 1993. Earthworm Resources and Vermiculture. Zoological Survey of India. Calcutta

Griffiths B. S. (2000): Ecosystem response of pasture soil community to fumigation induced microbial diversity reductions: Examination of the biodiversity-ecosystem function relationship. Oikos, 90: 279-294.

Hartenstein, R., Neuhauser, E.F. Kaplan D.L. (1979): Reproductive potential of the earthworm *Eisenia foetida*, Oecologia 43 (1979) 329–340.

Hendrix P. F. (1995): Earthworm ecology and biogeography in north America. Boca Raton, Fla.: Lewis Publishers.

Hertel K, Woodlands K (1998) 'Insect and mite control in field crops.' (NSW Agriculture: Orange, NSW)

House GJ (1985) Comparison of soil arthropods and earthworms from conventional and no-tillage agro-ecosystems. Soil and Tillage Research 5, 351-360. doi: 10.1016/S0167-1987(85)80003-9Isbel R (1996) 'The Australian Soil Classification.' Australian Soil and Land Survey Handbook. (CSIRO Publishing: Melbourne)

Ismail SA (1986) Earthworms resourses of Madaras. In: MC. Dash, B. K. Senapati and P. C. Mishra (eds) Worms and vermicomposting. Proc. Natl. Sem. Org. Waste Utilize. Vermicomp. Sambalpur University, Sambalpur, Orrisa, India pp. 8-15.

ISSS (1984): Soil science in India. Bull. No. 14. Indian soc of soil sci New Delhi

Jordan D, Stecker JA, Cacnio-Hubbard VN, Li F, Gantzer CJ, Brown JR (1997) Earthworm activity in no-tillage and conventional tillage systems in Missouri soils: A preliminary study. Soil Biology and Biochemistry 29, 489-491. doi: 10.1016/S0038-0717(96)00038-7

Ismail, S.A. 1997. Vermicology: Biology of Earthworm. Orient Longman, Hyderabad, India

Julka, J.M. 1993. Earthworm resources of India and their utilization in vermiculture. pp.51-56. In: Earthworm Resources and Vermiculture. Zoological Survey of India, Calcutta.

Kale RD and Krishnamoorthy RV (1981) Litter preferences in the earthworm *Lampito mauritii* Proc. India Acad. Sci. (Anim. Sci.)90: 123-128.

Keshvamoorthy p (1978). Seasonal migrations in earthworms in relation to soil composition. Ph. D. Thesis, submitted to the Banglore, India.

Killham, K (1994): Soil ecology Cambridge University Press New York

Kladivko E.J., Mackay A. D and Bradford J. M. (1986): earthworms as factor in the reduction of soil crusting. Soil sci. soc. Amar. J., 50: 191-196.

Kladivko EJ, Akhouri NM, Weesies G (1997) Earthworm populations and species distributions under no-till and conventional tillage in Indiana and Illinois. Soil Biology and Biochemistry 29, 613-615. doi: 10.1016/S0038-0717(96)00187-3

Lal R (1982) Effects often years of no-tillage and conventional plowing on maize yield and properties of a tropical soil. In 'Proceedings of the 9th Conference of the Soil Tillage Research Organization'. Osijek, Yugoslavia. pp. 111-117. (International Soil and Tillage Research Organisation)

Lavelle P. (1983): The soil fauna of tropical Savanas II. The earthworms. In: F. Bourliere (ed.) "Ecosystems of the world 13. Tropical Savans". Elsevier, Amsterdam. pp 485-504.

Lavelle P. (1997): Faunal activities and soil processes; adaptive strategies that determine ecosystem function. Advances in ecological research, 27/93-132.

Lee KE (1985) 'Earthworms: Their ecology and relationships with soils and land use.' (Academic Press: Sydney)

Madge D.S. (1969) Field and Laboratory studies on the activity of two species of tropical earthworms. Pedobiologia, 9: 188-214.

McCredie TA, Parker CA, Abbott I (1992) Population dynamics of the earthworm Aporrectodea trapezoides (Annelida: Lumbricidae) in a Western Australian pasture soil. Biology and Fertility of Soils 12, 285-289. doi: 10.1007/BF00336045

Michiels, N.K., Hohner, A. and Vorndran I.C. (2001): Precopulatory mate assessment in relation to body size in the earthworm *Lumbricus terrestris*: avoidance of dangerous liaisons?, Behav. Ecol. 12: 612–618.

Misra P. C. and Dash M. C. (1983): Population dynamics and production studies in earthworm population of an orison forest ecology. Trop. Ecol. (In Press)

Mohanjit (1986) Ecological studies on earthworms in relation to conservation o soil nutrients. Ph. D. Thesis. H.A.U., Hissar, India p. 238.

Monroy, F. Aira, M., Domínguez, J. Velando A. (2005): Size-assortative mating in the earthworm *Eisenia fetida* (Oligochaeta, Lumbricidae), J. Ethol. 23 (2005) 69–70.

Monroy, F., Aira, M. Velando, A. and Domínguez, J. (2003): Have spermatophores in *Eisenia fetida* (Oligochaeta, Lumbricidae) any reproductive role?, Pedobiologia 47 (2003) 526–529.

Neuhauser, F., D.L. Kaplan and R. Hartenstein. 1979. Life history of earthworm Eudrilus eugeniae. Review of Ecology and Biological Sciences 16: 525- 534.

NBSS (1995): soils of Maharashtra for optimizing land use. NBSS publ. 54. National Bureo of soil survey and land use planning., IARI campus New Delhi, India.

Nelson DW, Sommers LE (1982) Total carbon, organic carbon, and organic matter. In 'Methods of soil analysis. Part 2. Chemical and microbiological properties'. 2nd edn. (Ed. AL Page) pp. 539-579. (American Society of Agronomy: Madison, WI)

PAN Pesticide Database (2004) www.pesticideinfo.org/Index.html

PPI (1995): International soil fertility manual. Potash and phosphate Inst. Norcrosss, USA.

Rai M. M. (1995): principles of soil sci Macmillan India limited Darya Gang, New Delhi

Reinecke, A. J. and Kriel J.R. (1981): Influence of temperature on the reproduction of the earthworm *Eisenia foetida* (Oligochaeta), S. Afr.J. Zool. 16 (1981) 96–100.

Reinecke, A. J. and Venter, J.M. (1987): Moisture preferences, growth and reproduct0ion of the compost worm *Eisenia fetida* (Oligochaeta), Biol. Fertil. Soils 3 (1987) 135–141.

Rossi, J.P. (2003): Short-range structures in earthworm spatial distribution, Pedobiologia 47 (2003) 582–587.

Rozen, A (1988): The annual cycle in populations of earthworms (Lumbricidae Oligochaeta) in three types of oak-hornbeam of the Niepolomicka Forest II. Dynamics of population numbers, biomass and age structure, Pedobiologia 31 (1988) 169–178.

Sehgal J. (1996): pedology; concepts and applications. Kalyani publishers 1/1 Rajendra nagar Ludhiyna.

Sheppard, P. S. (1988): Specific differences in cocoon and hatchling production in *Eisenia fetida* and *Eisenia andrei*, in: C.A. Edwards, E.F. Neuhauser (Eds.), Earthworms in Waste and Environmental Management, SPB Academic Publishing BV, The Hague, The Netherlands, 1988, pp. 83–92.

Sims RW, EG Easton. (1972): A numerical revision of the earth-worm genus Pheretima auct. (Megascolecidae: Oligochaeta) with the recognition of new genera and an appendix on the earthworms collected by the Royal Society North Borneo Expedition. Biol. J. Linn. Soc. **4:** 169-268.

Stearns, S.C. (1992): The Evolution of Life Histories, Oxford University Press, Oxford, 1992.

Tembe V. V. and Dubash P. J. (1959): A preliminary note on the culture and development of Indian earthworm. J. Bomay. Nat. Hist. Soc., 56: 643-646.

Tiwari SC (1983) Effects of Organic manure and NPK fertilization on earthworm activity in Oxisol. Biol. Fertil. Soils 16: 293-295.

Tsukamoto, J. and Watanabe, H. (1976): Influence of temperature on hatching and growth of *Eisenia foetida* (Oligochaeta, Lumbricidae). Pedobiologia 17: 338–342.

Venter J.M., and Reinecke A. J. (1988): The life-cycle of the compost worm *Eisenia fetida* (Oligochaeta), S. Afr. J. Zool. 23 161–165.

Chapter 8

Karyo-Systematic Diversity and Evolution of the Cyprinids of Kashmir Valley

Farooq Ahmad Ganai, Farooq Ahmad Mir, Shahnawaz Ahmad and Yasir Irfan

P.G. Department of Zoology, Govt. Degree College Sopore, Kashmir, J&K

e-mail: farooqmd84@gmail.com, snahmad78@gmail.com

Cytogenetic study of eight cyprinids viz., Carassius carassius, Schizothorax niger, S. esocinus, S. curvifrons, S. plagiostomus, S. labiatus, Puntius conchonius, Crossocheilus diplocheilus was carried out by using smear splash technique. The diploid chromosome numbers deciphered were discussed in the purview of genetic evolution. Pattern of chromosomal evolution and role of chromosomal rearrangements was discussed in besides the cytotaxonomic status of these fishes.

Keywords: *Cytogenetics, Diversity, Evolution, Species, Karyotype.*

Introduction

Most animals have chromosome complements and karyotypes, which are characteristic for the species to which they belong. A karyotype is numerical and physical attributes of chromosomes of a eukaryotic cell (Stebbins, 1950; White, 1973). In Karyological investigations, keen consideration is given to chromosomal length, Centromeric position and allosomic differentiation and such studies have become an indispensable in addressing taxonomic issues. The chromosomes se displayed in pairs an order based on length and Centromeric position. Without a complete knowledge of karyotypes of plants and animals, it is impossible to establish the exact evolutionary relationship between them. The cytogenetic studies of fish have become a vibrant domain of cytologyasthis vertebrate group is mostspecies rich group, with 30000 known species. The fish present a perplexing diversity in biological paradigm Cytogenetic investigations have also lent details about chromosome morphology which is necessary for carrying out genetic manipulation in fish. Karyotypic analysis is mandatory for *ex situ* fish breeding in purview of genetic control, instant generation of true lines and evolutionary pursuits. Cytogenetic investigations are paramount to aquacultural practices in relation to genetic manipulations like production of polyploids, gynogenesis, androgenesis and generation of hybrids. Karyological studies help understand the changes that ramified an ancestral karyotype in the course of evolution.

Kashmir Himalaya harbours a number of aquatic habitats inhabited by different type of fish. The work on fish of Kashmir was started by Heckel (1838) who for the first time reported about sixteen species of fish in the valley, all new to science. Since that time many others have worked on the fish and fishery of Kashmir with the work of Mukerji (1936), Hora (1936), Silas (1960), Das and Subla (1963), Qadri *et al.* (1983), Yousuf (1996), Kullander *et al.* (1999) as the prominent ones. Although creditable work has been done by the earlier workers in identifying and allotting taxonomic position to the fishof Kashmir, the work is mostly based on morphometric and meristic characters. Due to plasticity of morphological characters there is a lot of confusion regarding the taxonomic positioning of cyprinid fishes of the valley. The possibility of hybrids (Heckel, 1838; Hora, 1936; Mukerjee, 1936) has further complicated the taxonomic identity of different species. In order to resolve the confusion existing in the taxonomic status of different schizothoracine fish species of the valley, it was decided to take the help of cytogenetics in the current study on some eight fish species (five schizothoracines and three cyprinides) of Kashmir valley, to unravel their cytogenetic make up at molecular level and expose their path of evolution and cytotaxomic position. Therefore, the eight fish species *viz. Carassius carassius, Schizothorax labiatus, S. plagiostomus, S. niger, S. curvifrons, S. esocinus, Crossocheilus diplochilus* and *Puntius conchonius* were selected for the present research to provide the insight and confirmation of karotypic diversity which would help in future concerned studies.

Materials and Methodology

1. Collection of Fish

Live fish samples were obtained from the local fish markets and were brought to the laboratory and acclimatized in a well-ventilated aquarium.

2. Cytogenetic Analysis

Modified air dried chromosome preparation method (Thorgaard and Disney,

1990) was used for the cytogenetic analysis of the 48 fish specimens (six of each species) which received two doses of phytohemagglutinin (PHA) injections (4μgg^{-1} BW), in a 20-h time interval at 20°C. Eight hours after the second PHA injection, 0.05 per cent colchicine was injected intraperitoneally @ 0.1ml/100g BW to depress the mitotic division at the metaphase stage and left for 2-3 hours before sacrificing. The fish were anesthetized by 300 ppm clove oil for 40s, their anterior kidney was removed, homogenized and hypotonised simultaneously by 0.56 per cent potassium chloride for 35 minutes at room temperature. Suspensions were centrifuged at 1000 rpm for 10 minutes. Supernatant was removed and the cells were fixed by cold fresh Carnoy fixative (3:1 methanol and glacial acetic acid). This fixation process was repeated three times and the cold fresh Carnoy was replaced at 30 minute time intervals. Smears were prepared on cold lamellae using splash method from 1m height, air dried for 24 h and stained with 2 per cent Giemsa.

3. Chromosomal Analysis

Leica DM LS2 trinoccular photomicroscope with 1000X magnification lens (oil immersion) was used for the photographic analysis of the chromosomes. Eighty metaphase plates were counted per species and a proper plate was selected to obtain the karyotype formulae. For each chromosome centromeric index, arm ratio and total length was calculated as described by Levan *et al.* (1964). The fundamental arm number was also calculated and the chromosomes were classified into metacentric, sub-metacentric, sub-telocentric and telocentric following the method of Levan *et al.* (1964).

Results and Discussion

An accurate knowledge of karyotypes of plants and animals is necessary for establishing the exact evolutionary relationship between them (MacGregor, 1993; Winkler *et al.*, 2004). Karyotype studies have become increasingly valuable for solving systematic and evolutionary problems in fish. Generally, species with high chromosome numbers are considered to have resulted through polyploidy from ancestral diploids with 2n equalling 48 or 50 (Rishi *et al.*, 1998). During the present study eight species of Cyprinids including *Carassius carassius, Schizothorax niger, S. esocinus, S. curvifrons, S. Plagiostomus* and, *S. labiatus, Puntius conchonius, Crossocheilus diplocheilus* were studied for different karyological features. Of all the fish species assessed cytogenetically in the present study, *Crossocheilus diplochilus* and *Puntius conchonius* revealed typical diploid number of chromosomes (2n=50) as both of them are smaller in size and therefore seem to be the ancestral diploids. However, the remaining six species are significantly larger so clearly exhibited polyploidy (chromosome number 94–100). *Carassius carassius*whose diploid chromosome number (Figure 8.1) was 2n=100 (80 per cent investigated metaphases) wherein the karyotype consisted of 12 pairs of metacentric, 2 pairs of submetacentric, and 36 pairs of telocentric chromosomes with a fundamental arm number (FN) of 128 is still poorly understood, since only a few studies based on Giemsa-stained chromosomes are known to be available about it. The number

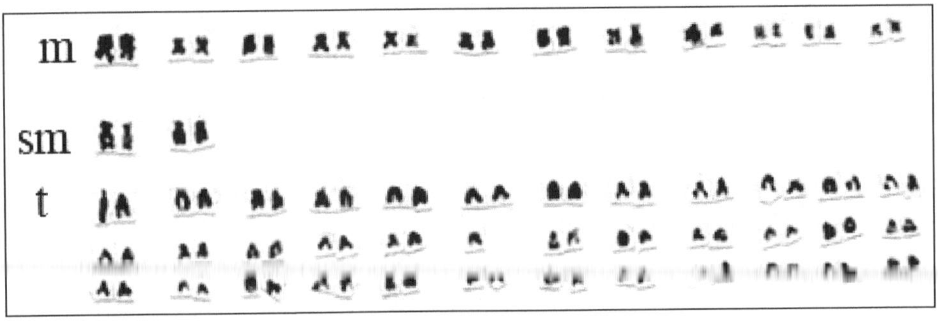

Figure 8.1: Karyotype of *C. carassius*.

however equalled the value reported in all previous studies except those by Raicu *et al.* (1981) and Chiarelli *et al.* (1969). At present, it is difficult to speculate more about the observed difference between the reported chromosome numbers unless detailed population screening of this species is available. Contrasting to the results obtained by Raicu *et al.* (1981), the diploid number of 104 chromosomes presented by Chiarelli *et al.* (1969) could be most likely attributed to preparation artefact usually created during the splashing of the cell suspension. Differences in the karyomorphology can be attributed to uneven chromosome spiralization and differential arm contraction induced by the pre-treatment (Farooq *et al.*, 2012).

Although the three species of *Schizothorax (niger*, *esocinus*, and *labiatus*) revealed a diploid number of 98 chromosomes with fundamental arm number of FN=154, 152 and 142 respectively, their chromosome combination was different with *S. niger* revealing a combination of 12 metacentric pairs, 16 submetacentric pairs, 11 subtelocentric pairs and 10 telocentric pairs (Figure 8.2), *S. esocinus* revealing a combination of 15 metacentric chromosome pairs, 11 submetacentric pairs, 5 subtelocentric pairs and 18 telocentric pairs (Figure 8.3) and *S. labiatus* revealing a combination of 12 metacentric pairs, 10 submetacentric pairs, 1 subtelocentric pair and 26 telocentric pairs (Figure 8.4) of chromosomes. The variation in the fundamental number without change in the 2n may be attributed to the

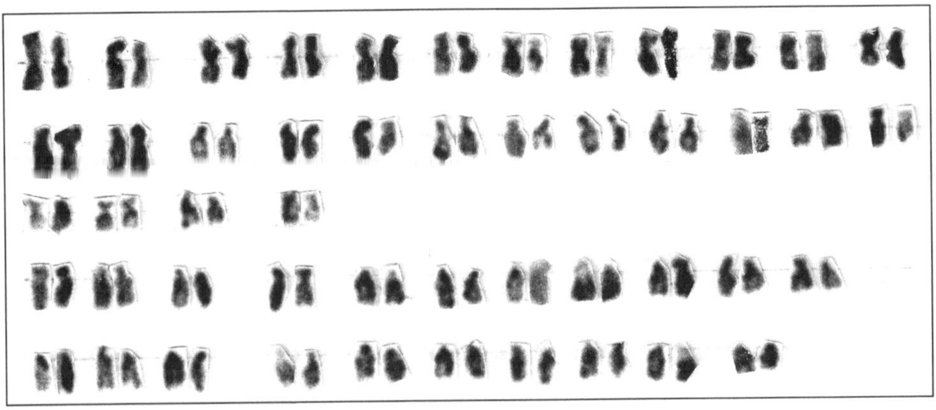

Figure 8.2: Karyotype of *S. niger*.

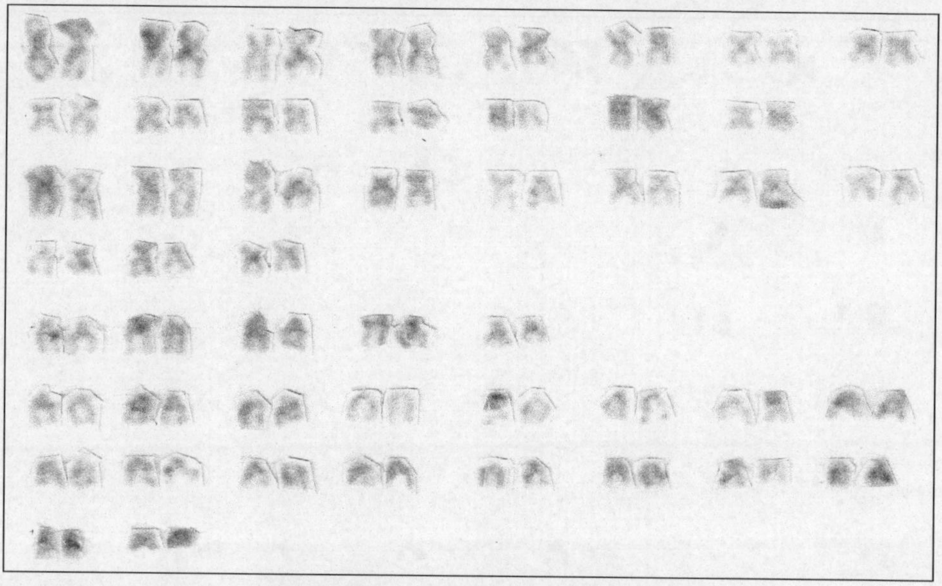

Figure 8.3: Karyotype of *S. esocinus.*

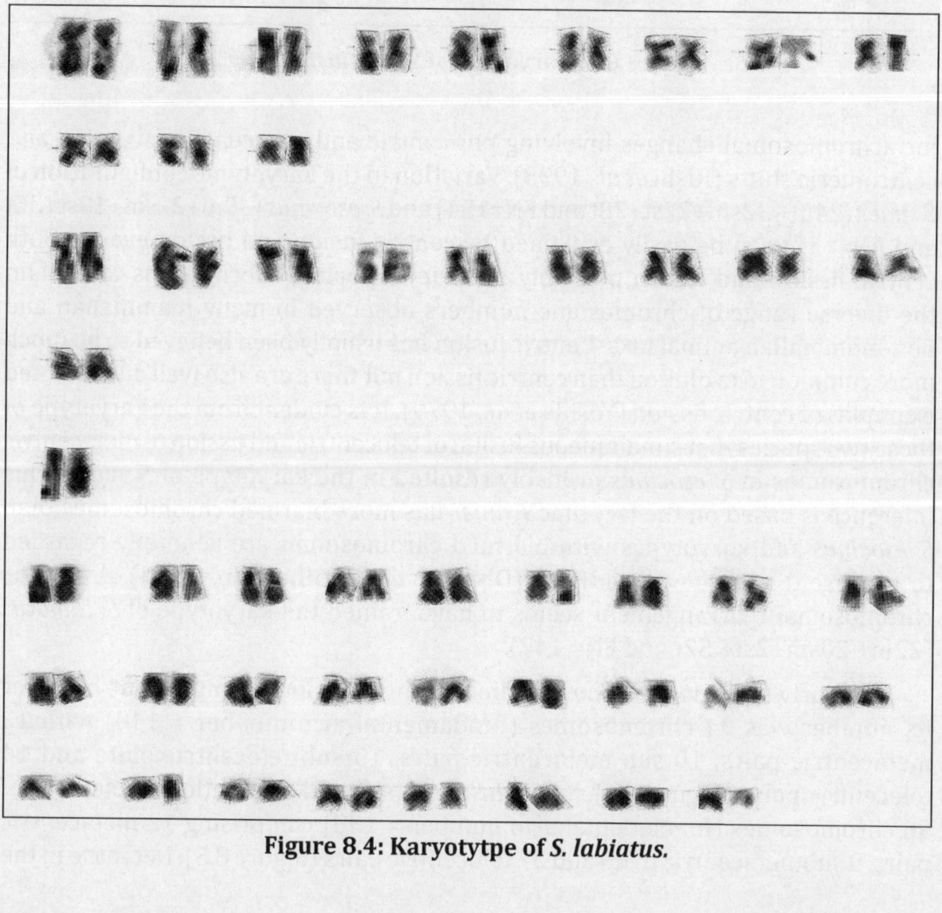

Figure 8.4: Karyotytpe of *S. labiatus.*

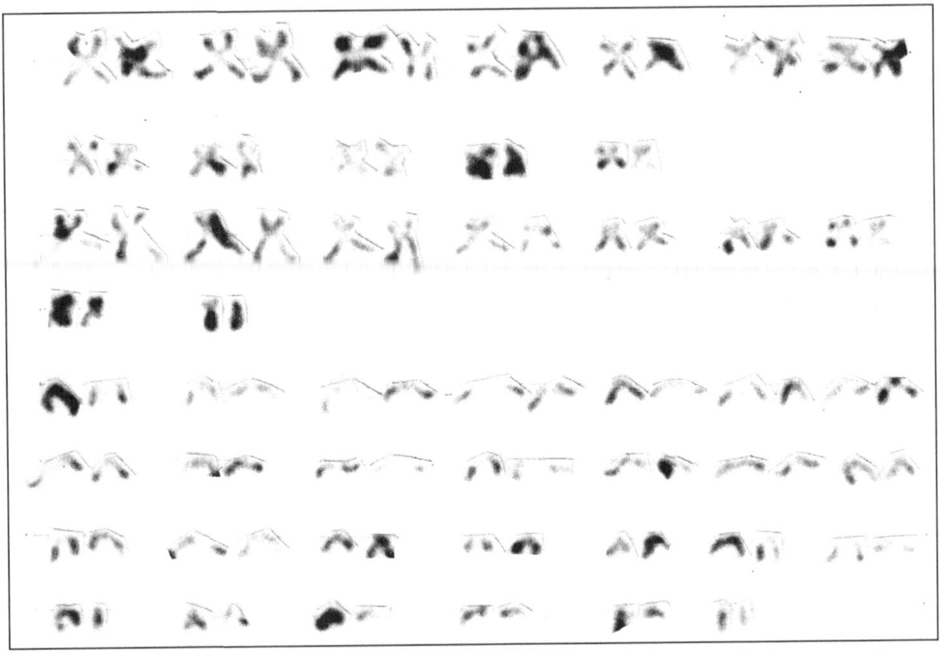

Figure 8.5: Karyotype of *S. plagiostomus*.

intrachromosomal changes involving pericentric and paracentric inversion and centromeric shifts (Rishi *et al.*, 1998). Variation in the karyotypic configuration of *S. niger* (24m+ 32sm+22st+20t and FN=154) and *S. esocinus* (30m+22sm+10st+36t and FN=152) can be easily explained by centric fusion and fission events. Both centric fission and fusion probably provide important mechanisms to explain the diverse range of chromosome numbers observed in many mammalian and non-mammalian animal taxa. Centric fusion has usually been believed to be much more common in evolution than centric fission but there are also well documented examples of centric fission (Yosida *et al.*, 1979). It is evident from the karyotype of these two species that simultaneous fusion of telocentric and fission of metacentric chromosomes in *S. esocinus* probably resulted in the karyotype of *S. niger*. This inference is based on the fact that *S. niger* has more biarmed chromosomes than *S. esocinus* and karyotypes with biarmed chromosomes are generally regarded to represent a derived condition (Ohno *et al.*, 1968; Ohno, 1970). A similar chromosomal rearrangement seems to have framed the karyotype of *S. labiatus* (22m+ 20sm+2st+ 52t and FN= 142).

Similarly, *S. curvifrons* showed diploid chromosomal complement however its number was 94 chromosomes (fundamental arm number = 140) with 13 metacentric pairs, 10 sub metacentric pairs, 10 sub telocentric pairs and 14 telocentric pairs (Figure 8.6). *S. plagiostomus* showed a somatic complement of 96 chromosomes (fundamental arm number = 138) comprising 12 metacentric pairs, 9 submetacentric pairs and 27 telocentric pairs (Figure 8.5). Decrease in the

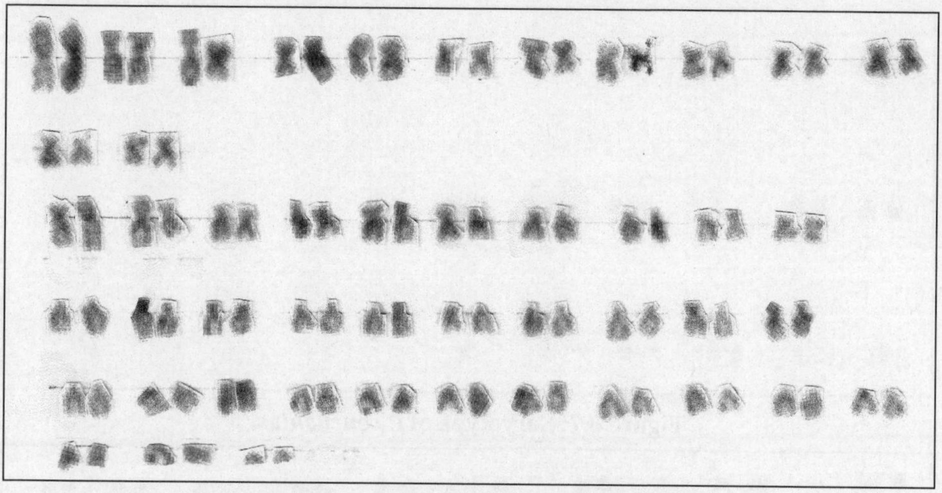

Figure 8.6: Karyotype of *S. curvifrons*.

2n and FN in these species may be attributed to the Robertsonian arrangements and pericentric inversion.

Both *Crossocheilus diplochelus* and *Puntius conchonius* showed a diploid number of chromosomes (2n=50). The diploid chromosome number of *P. conchonius* (Figure 8.7) consisted of 11 pairs of metacentric, 8 pairs of submetacentric and 6 pairs of telocentric chromosomes respectively with the fundamental arm number (FN) equalling 88. The diploid chromosomal complement of *Crossocheilus diplochelus* consisted of 50 chromosomes (Figure 8.8), comprising 9 metacentric pairs and 16 telocentric pairs with the fundamental arm number (FN) equalling 68. The diploid chromosome number (50) is the most commonly occurring number in family cyprinidae and is considered as the modal number in case of this family. Though the two species cannot be compared in chromosomal structure for belonging to two different genera but owing to the fact that both are members of the same family, it can be inferred that *C diplochilus* is more primitive than *P conchonius* as the latter exhibits a derived karyotype. Moreover, according to the studies performed by various workers on *Puntius* species (Nayyar, 1964), it seems that 2n equals to 50 in the genus *Puntius,* as in many other cyprinids and the present study also revealed the same diploid number. Despite the similarity of the diploid number in different species of *Puntius*, there are differences in their karyotype formulaeal though Nayyar (1964) reported the presence of all acrocentric chromosomes in *P. conchonius*, both biarmed and acrocentric chromosomes were observed in this species during the present study. The primitive teleost karyotype is thought to have consisted of 46 to 48 acrocentrics (Nayyar, 1966; Ohno *et al.*, 1968) and Karyotypes with biramous chromosomes are considered evolutionarily advanced (Ohno, 1970). *P. conchonius* collected from Dal Lake showed a derived karyotype configuration

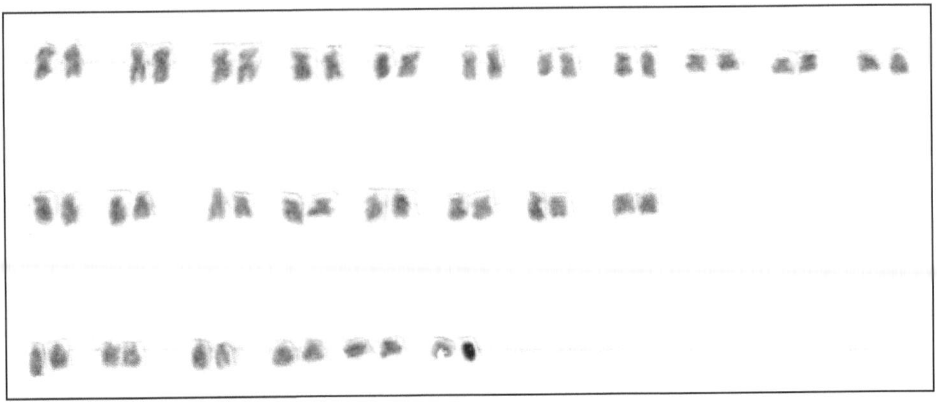

Figure 8.7: Karyotype of *P. conchonius*.

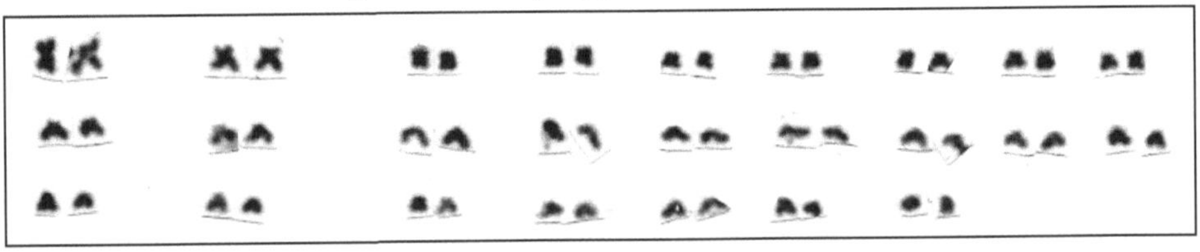

Figure 8.8: Karyotype of *C. diplochelus*.

and it appears to be a cytotype of *P. conchonius* found in Jammu region of the State as that species has only 48 chromosomes.

The chromosomes of *S. niger*, *S. esocinus*, *S. labiatus* and *S. curvifrons* were categorised into four groups *i.e.*, metacentric, submetacentric, subtelocentric and telocentric following Levan *et al.* (1964). In *S. plagiostomus*, *C. carassius*, *C. diplochilus* and *P. conchonius* none of the chromosomes could be recognised as sub telocentric owing to the presence of relatively small chromosomes with their centromeric positions ranging gradually from median to nearly terminal, making it difficult to assign some chromosomes to a particular chromosomal category and thus making correct identification of individual chromosomes nearly impossible. The karyological study of teleost fish presents technical difficulties which are not encountered in the study of other vertebrates and these difficulties are due to small size and high number of chromosomes (Cucchi and Baruffaldi, 1990). Further, differential arm contraction can alter a chromosome classification in a karyotypic formula.

The maximumt number of biramous chromosomes were seen in *S. niger* followed by *S. esocinus*, *S. curvifrons*, *S. labiatus*, *S. plagiostomus*, *P. conchonius*, *C. carassius* and *C. diplochilus,* whereas the maximum number of subtelo-telocentrics were observed in the order of *C. carassius* > *S. Plagiostomus* > *S. Labiatus* > *S. curvifrons* > *S. Esocinus* > *S. niger*. The ancestral teleost karyotype is opined to have possessed 46–48 acrocentrics and the karyotypes with two armed condition is a latest development. Keeping this into consideration *C. carassius*, *C. diplochilus*,

P. conchonius, *S. plagiostomus* and *S. labiatus* seem to be primitive fishes when compared to *S. niger*, *S. esocinus* and *S. curvifrons* which possess a more derived karyotype. The trend of gradual increase in the FN from 68 in *C. diplochilus* to 154 in *S. niger* supports the above inference and hence, the family cyprinidae has undergone all the possible kinds of chromosomal mix ups for better karyological evolution and diversity.

Therefore, it is pertinent to mention that some species of *Schizothorax* exhibited karyotypes similar to those recorded for other species inhabiting distant geographical locations, *e.g.*, *Schizothorax richardsonii* (2n=98), *Schizothoraichthys progsastus* (2n=98) and *S. kumaonensis* (2n=98) suggesting their origin from the common ancestor. This close proximity in cytogenetic makeup lends support to the notion that *Schizothorax* species are very precisely similar in that they have not been isolated as evolving entities long enough for random chromosome changes to have taken place and become fixed.

From our study and in the light of available cytological literature along the hypothetical model of Nakatani *et al.* (2007), we propose the origin of Kashmir *Schizothorax* from *S. richardsonii,* which seems more plausible and with reason on account of its close resemblance in genetic as well as morphological features. We further propose that *S. richardsonii* arose either from *S. zarudnyi* (Iran) or some other species, which warrants a meticulous cytogenetic study of all the extent fishes of this group. And it is *S. richardsonii* which most likely gave rise to *S. plagiostomus* through such mechanisms as are important for change in karyomorpholgy and 2n, *viz.*, Robertsonian arrangements, fissions and centric fusion. Genetic changes lead to chromosomal incompatibilities and reproductive isolation which lead to further diversification and evolution. *S. plagiostomus* explored new aquatic habitats, encountered new changes and continued evolving. This evolution continued spatio-temporaly and resulted in the generation of five *Schizothorax* species of Kashmir. This hypothesis is supported by the karyomorpholgy and other cytogenetic markers of fish (Table 8.1). Moreover, researchers all-round the globe have put their efforts to trace out the teleost ancestor, path of its evolution and off shoots thereof. Increase in the genome size can be attributed to the process of whole genome duplications and there have been several rounds of whole genome duplications in the course of evolution of vertebrates. Similarly, other Kashmir schizothoracines like *Diptychys* and *Ptychobarbus* inhabiting the high waters of Ladakh and Tibetian plateau also witnessed an additional round of genome duplication which could be attributed to the possible role of geographical isolation as an evolutionary tool. When the Himalaya was originating and the Tibetian plateau was gaining height, the ichthyofauna of Kashmir valley separated from its counterparts in aquatic habitats of Tibet and allied regions. The overall karyotypic evolution of the presently studied cyprinids can be understood from the given dendrogram (Figure 8.9) which is based on the cytogenetic profile of these fish. It is clear from the dendrogram that they come from some common teleost ancestor and after the 3R-WGD, their ancestral genome experienced large-scale chromosomal rearrangements, including fissions,

Table 8.1: Probable Source of Origin of *Schizothorax* Species

S. No.	Name of the species	2n	Chromosome morphology				NF value	Author and Year
			m	Sm	St	t		
1.	*S. kumaonensis*	98	24	6	68		128	Rishi et al., 1998
2.	*S. kumaonensis*	98	18		70	10	126	Lakara et al 1997
3.	*S. progastus*	98	16	20	12	50	134	Rishi et al., 1983
4	*S. richardsoni*	98	16		42	40	154	Lakara et al., 1997
5.	*S. prenanti*	148	28	40	36	44	216	Yu et al., 1987
6	*S. zarudnyi*	96	18	28		50	142	Kalbassi et al.2008
7.	*S. plagiostomus*	96	24	18		54	138	Present work
8.	*S. labiatus*	98	24	20	2	52	142	Present work
9.	*S. curvifrons*	94	26	20	20	28	140	Present work
10.	*S. esocinus*	98	30	22	10	36	152	Present work
11.	*S. niger*	98	24	32	22	20	154	Present work

fusions, and translocations which shaped their cytogenetic make-up. It is evident from the dendrogram that *S. labiatus, S. plagiostomus* have evolved along the same evolutionary line while *S. labiatus* and *S. plagiostomus* are more related to each other than the other cyprinids as supported by the studied cytogenetic parameters. These two fish are karyologically primitive and that is the reason they represent a single cluster and have diversified late in the evolutionary time scale. Form another evolutionary cluster of *S. esocinus, S. curvifrons* and *S. niger* the first two seems to be more closely related to each other than third one and at the base all the three share a common evolutionary history with some diploid ancestor, suggesting that these cyprinids have originated from a common ancestor with a 2n equalling 48–50. The 3R-WGD is expected to have occurred in the teleost ancestor, but after its split from living non-teleost actinopterygians based on analyses of the number of loci

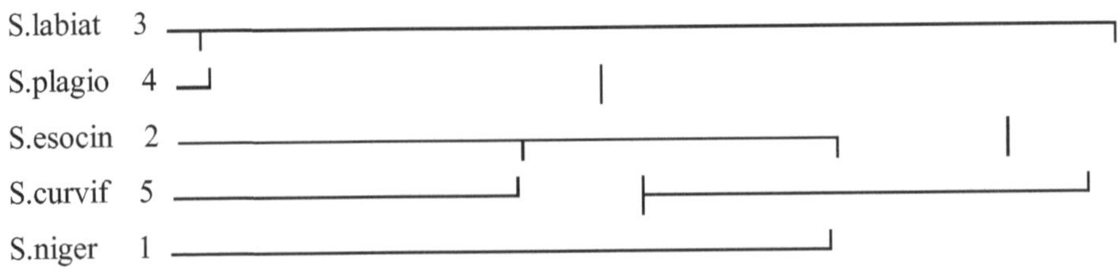

Figure 8.9: Phylogenetic Relationship of *Schizothorax* of Kashmir.

and phylogenies of several nuclear genes (Sato and Nishida 2007; Chambers *et al.*, 2009). Further, the dendrogram supports the endemic nature of *S. niger*, as it occupies the lowest branch of the dendrogram among polyploid cyprinids, infering that it has evolved from *S. esocinus* and represents the culminated *Schizothorax* species of this evolutionary lineage.

The general and overall evolutionary relationship of Kashmir valley cyprinids can be understood from Figure 8.10 as it is clear from the figure that *C. carassius* is karyo-evolutionarily more related and close to *S. labiatus* and *S. plagiostomus* than any other species. Since *C. carasius* is an introduced species and it can be regarded as a co-evolutionary cyprinid species. Similarly, *P. conchonius* and *C. diplochilus* occupy the base of the dendrogram which lends support to the suggestion of the origin of tetra and hexaploids from the diploid ancestors.

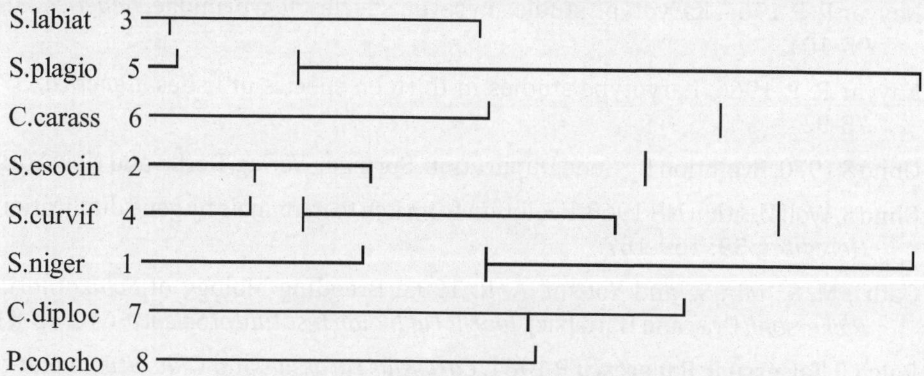

Figure 8.10: Phylogenetic Relationship of Cyprinids of Kashmir.

REFERENCES

Chiarelli B, Ferrantelli O, Cucchi C 1969. The karyotype of some teleostean fish obtained by tissues culture *in vitro*. *Experientia*, 25(4): 426–427. doi: 10.1007/BF01899963.

Cucchi, C and Baruffaldi, A., 1990. A new method for karyological studies in teleost fishes. *J. Fish. Biol.* 37: 71-75.

Das, S. M., and Subla, B. A. 1963. The icthyofauna of Kashmir: Part 1. "History, topography, origin, ecology and general distribution". *Icthyologica*, 2: 87-106.

Farooq A.G, Yousuf A.R, Sabzar A.D, Tripathi N.K and Samee wani., 2012. Cytotaxonomic status of Schizothoracine fishes of Kashmir Himalaya (Teleostei: Cyprinidae).*Caryologia*,

Heckel, J. J., 1838. FischeausCaschmir. Carl Freiherrn V. Hugel, Wien.

Hora, S. L., 1936. Report on fishes. Part 1: Cobitidae. *Memoirs of the Connecticut Academy of Arts and Sciences*, **10:** 299-321.

Kullander, S.O., Fang, F., Delling, B and Ahlander, E., 1999. The fishes of Kashmir Valley. *In*: River Jhelum, Kashmir Valley, Impacts on the aquatic environment. P. 99-163. Lenart Nyman Ed.

Levan, A., Fredga, K and Sandberg, A. A., 1964. A nomenclature for centromeric position on chromosomes. *Heriditas*, 52: 201-220.

MacGregor, U.C., 1993. Chromosome preparation and analysis. Chapter 6: 177-186.

Mukerji, D. D. 1936. Report on fishes. Part II: *Sisoridae, Cyprinidae. Memoirs of the Connecticut Academy of Arts and Sciences*, 10: 323-359.

Nakatani Y, Takeda H, Kohara Y, Morishita S., 2007. Reconstruction of the vertebrate ancestral genome reveals dynamic genome reorganization in early vertebrates. Genome Res 17: 1254–1265. doi: 10.1101/gr.6316407

Nayyar R. P. 1964. Karyotype studies in seven species of cyprinidae. *Genetica*, **35:** 95-104.

Nayyar R. P. 1966. Karyotype studies in thirteen species of fishes. *Genetica*, **37:** 78-92.

Ohno S 1970. Evolution By Gene Duplication. Springer-Verlag, Berlin and New York.

Ohno S, Wolf U, Atkin NB 1968. Evolution from fish to mammals by gene duplication. *Hereditas*, **59:** 169-187.

Qadri, M. S., Mir, S., and Yousuf, A. R. 1983. Breeding biology of *Schizothorax Richarsonii* Gray and Hard [sic]. *Journal of Indian Institute of Science*, 64C: 73-81.

Raicu P, Taisescu E, Banarescu P 1981. *Carassius carassius* and *C. auratus*, a pair of diploid and tetraploid representative species (Pisces, Cyprinidae). *Genetica*, **46:** 233–240.

Rishi, K. K., Shashikala and Rishi, S., 1998. Karyotype study on six Indian hill-stream fishes. *Chromosome Science*, **2**: 9-13.

Sato Y, Nishida M 2007.Teleost fish with specific genome duplication as unique models of vertebrate evolution.Environ. Biol. Fish. 88: 169–188.DOI 10.1007/s10641-010-9628-7

Silas, E.G. 1960. Fish from the Kashmir Valley. *Journal of the Bombay Natural History* Society, 57: 66-77.

Stebbins, G. L., 1950. Variation and evolution in plants. Chap. XII. The karyotype. Columbia University Press, N.Y.

Thorgaard, G. H and Disney, J. E., 1990. Chromosome preparation and analysis. In: Schreck C.B, Moyle P.B, eds. '*Methods for fish biology*'. Bethesda, M.D: American Fisheries Society.pp.171-190.

White, M. J. D., 1973. The chromosomes. 6[th.] Ed. Chapman and Hall, London.

Winkler, F. M., Garcia-Melys, D and Palma-Rojas, C., 2004. Karyotypes of three South East Pacific Flounder species of the family Paralichthyidae. *Aquaculture Research*. 35: 1295-1298.

Yosida, T. H., Kato, H., Tsuchiya, K., Moriwaki, K., Ochiai, Y and Monty, J., 1979. Mauritius type black rats with peculiar karyotypes derived from Robertsonian fission of small metacentrics. *Chromosoma*, 75: 51-62.

Yousuf, A. R. 1996. Fishery resources of Kashmir. *In: Ecology, Environment and Energy. Eds*. A. H. Khan and A. K. Pandit. University of Kashmir, Srinagar. MS only, 20 p.

Chapter 9

Comparative Study of Seasonal Changes of the Protein in the Tissues of Freshwater Male Catfish *Mystus cavasius* (Ham) from Bhadra River and Bhadra Reservoir, Karnataka State, India

H.M. Ashashree, H.A. Sayeswara*, A.R. Shantha and K.L. Naik*

Department of Zoology, Sahyadri Science College (Autonomous), Shivamogga, Karnataka
**Corresponding authors e-mail: ashashree2013@gmail.com; sayesh2009@gmail.com*

Biochemical constituents of the body of the freshwater fish Mystus cavasius were greatly influenced by the breeding activity. Fish is one of the main natural sources of protein and provide certain other useful products as well as economic substance to many countries. Now a day fish has to become a healthier alternative food meat for the last sixty years. Seasonal variations in the various biochemical constituents of muscle, liver and gonad were studied in the catfish Mystus cavasius. The protein content of the muscle and liver showed peak periods in post-breeding and spent seasons. There was an increase in protein levels of testes during breeding seasons. The seasonal cycles of the fish seemed to be governed by different environmental factors.

Keywords: *Mystus cavasius, Bhadra river, Seasons, Tissues, Biochemical.*

Introduction

Biochemical compositions of fish tissue are of considerable interest for their specificity in relation to food values of fish and for evaluating their physiological needs at different periods of life (Kiran, 2018). Protein is the most important and complex group of biological materials, as the form the chief nitrogenous constituents of the tissues of the body. Many of the fishes exhibit considerable seasonal changes in their composition due to changes in the environmental due to changes in the environmental, conditions such as availability of nutrients, pollution of water and fluctuations in physico-chemical parameters and hydrographic features. Biochemical assessment is necessary to ensure the nutritional value as well as palatability of the fresh fish. (Azam *et al.,* 2004). Bio-indices have been developed to provide the measurement of general condition of fish population and indicate the physiological status of the individual fish. *Mystus cavasius* is a commercially important fish having high protein contents and taste. The seasonal biochemical variations correlating testes muscle and live have not been given much attention of *Mystus canvsius* of Bhadra River. Hence, keeping in view of the commercial importance of fish, an attempt have been made to study the storage and utilization of liver contents for gonadal growths.

Materials and Methods

The male fish *Mystus cavasius* were collected several times per month from Bhadra river during January 2018 to December 2018 and were obtained from fisherman. They were brought in to the laboratory and then scarified for further studies. The tissue was processed for protein estimations. The protein was estimated as method described by Lowry *et al.* (1951). Statistical Analysis: The values are expressed as mean ± standard deviation (SD). The results were evaluated by using the SPSS (version 12.0) and Origin 6 software.

Results and Discussion

The results of the analysis of biochemical compositions of the present fish are given in the Tables 9.1–9.5 and depicted in Figures 9.1–9.27. From the study it was concluded that the *Mystus cavasius* is an excellent source of protein because of its low fat content and it is one of the best nutrients foods for human body due to its abundant protein. The gonad protein is highest during monsoon season and lowest during spent season. The muscle and liver protein starts decline in monsoon season and highest in post breeding and spent seasons.

The biochemical compositions of the whole body indicates quality of the fish, therefore biochemical composition of a species helps to assess its nutritional and edible value in terms of energy units compared to other species, the biochemical contents depends on fishing ground, season age, sex, and reproductive status of the individual. A variety of factors such as food, space, temperature, salinity, physical activity influence the growth of fish (Weatherly; Ahmed *et al.,* 2012) and the fish body elements may change due to these factors (Kamal *et al.,* 2007). The

Table 9.1: Comparison of Monthly Variations of Proteins in Testis of *Mystus cavasius* (HAM) in Bhadra Reservoir and Bhadra River

Months	*Bhadra Reservoir*	*Bhadra River*
January	1.20±0.01	0.94±0.01
February	1.43±0.02	1.26±0.02
March	1.62±0.01	1.45±0.01
April	1.68±0.12	1.56±0.09
May	1.71±0.06	1.59±0.04
June	1.89±0.06	1.70±0.02
July	1.99±0.04	1.98±0.03
August	1.77±0.03	1.02±0.02
September	1.66±0.08	1.56±0.01
October	1.85±0.01	1.85±0.07
November	1.27±0.01	0.77±0.05
December	0.99±0.02	0.63±0.06

Table 9.2: Comparison of Monthly Variations of Proteins in Liver of *Mystus cavasius* (HAM) in Bhadra Reservoir and Bhadra River

Months	*Bhadra Reservoir*	*Bhadra River*
January	9.11±0.01	8.15±0.02
February	6.23±0.03	6.50±0.02
March	5.03±0.01	5.52±0.06
April	4.02±0.13	3.88±0.09
May	2.95±0.09	2.11±0.07
June	1.21±0.06	1.11±0.05
July	1.99±0.04	1.38±0.02
August	3.95±0.01	2.02±0.03
September	4.66±0.05	3.56±0.09
October	8.21±0.06	6.85±0.01
November	11.01±0.01	8.77±0.01
December	13.99±0.02	9.63±0.04

spawning cycle and food supply are the main factors responsible for this variation (Love, 1980). The relationship between GSI and HIS was found to be negatively significant throughout the study period as well as during the progression of gonad maturation stages. Monthly variations of protein contents of liver with testes shows an inverses relationship *i.e.*, in the testicular protein reaching maximum during July (2.99±0.01), on the contrary the hepatic protein reaching minimum during July (1.20±0.02) suggests the involvement of hepatic protein for testicular growth.

Table 9.3: Comparison of Monthly Variations of Proteins in Muscles of *Mystus cavasius* (HAM) in Bhadra Reservoir and Bhadra River

Months	*Bhadra Reservoir*	*Bhadra River*
January	5.85±0.01	5.62±0.11
February	3.93±0.02	2.99±0.03
March	2.62±0.01	2.25±0.09
April	1.87±0.12	1.85±0.10
May	1.61±0.06	1.57±0.05
June	0.89±0.06	0.70±0.07
July	0.59±0.04	0.58±0.01
August	1.27±0.03	1.12±0.01
September	1.46±0.08	1.39±0.03
October	1.95±0.01	1.85±0.06
November	2.27±0.01	2.13±0.09
December	4.99±0.02	4.63±0.03

Table 9.4: Monthly Variations of Proteins in different Organs of *Mystus cavasius* (HAM) in Bhadra Reservoir

Months	*Testis*	*Liver*	*Muscle*
January	1.20±0.01	9.11±0.01	5.85±0.01
February	1.43±0.02	6.23±0.03	3.93±0.02
March	1.62±0.01	5.03±0.01	2.62±0.01
April	1.68±0.12	4.02±0.13	1.87±0.12
May	1.71±0.06	2.95±0.09	1.61±0.06
June	1.89±0.06	1.21±0.06	0.89±0.06
July	1.99±0.04	1.99±0.04	0.59±0.04
August	1.77±0.03	3.95±0.01	1.27±0.03
September	1.66±0.08	4.66±0.05	1.46±0.08
October	1.85±0.01	8.21±0.06	1.95±0.01
November	1.27±0.01	11.01±0.01	2.27±0.01
December	0.99±0.02	13.99±0.02	4.99±0.02

Proteins play an important role in enzymes catalysis in transport or mechanical support for the control of growth and maturation of testes. Comparison of fish protein between Bhadra reservoir and Bhadra River is summarized in Tables 9.1–9.5. High proteins were observed from Bhadra Reservoir compared to Bhadra River because of physico-chemical parameters and domestic wastes. There are factors responsible for fish retardation other than physical stability like pollution

Table 9.5: Monthly Variations of Proteins in different Organs of Bhadra River

Months	*Testis*	*Liver*	*Muscle*
January	0.94±0.01	8.15±0.02	5.62±0.11
February	1.26±0.02	6.50±0.02	2.99±0.03
March	1.45±0.01	5.52±0.06	2.25±0.09
April	1.56±0.09	3.88±0.09	1.85±0.10
May	1.59±0.04	2.11±0.07	1.57±0.05
June	1.70±0.02	1.11±0.05	0.70±0.07
July	1.98±0.03	1.38±0.02	0.58±0.01
August	1.02±0.02	2.02±0.03	1.12±0.01
September	1.56±0.01	3.56±0.09	1.39±0.03
October	1.85±0.07	6.85±0.01	1.85±0.06
November	0.77±0.05	8.77±0.01	2.13±0.09
December	0.63±0.06	9.63±0.04	4.63±0.03

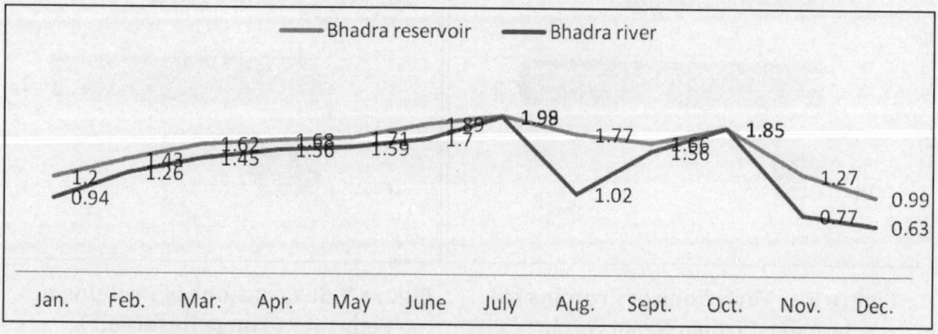

Figure 9.1: Comparison of Monthly Variations of Proteins in Testis of *Mystus cavasius* of Bhadra Reservoir and Bhadra River.

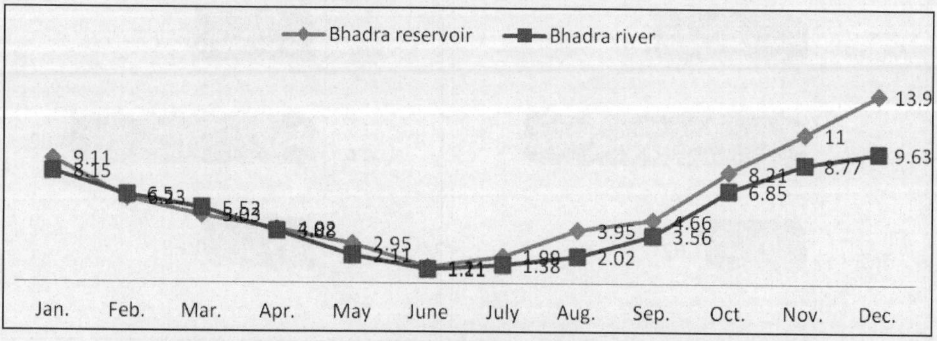

Figure 9.2: Comparison of Monthly Variations of Proteins in Liver of *Mystus cavasius* of Bhadra Reservoir and Bhadra River.

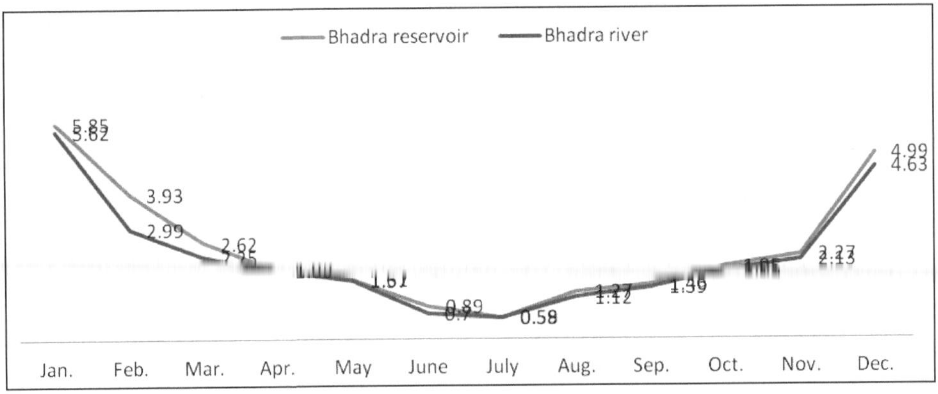

Figure 9.3: Comparison of Monthly Variations of Proteins in Muscles of *Mystus cavasius* of Bhadra Reservoir and Bhadra River.

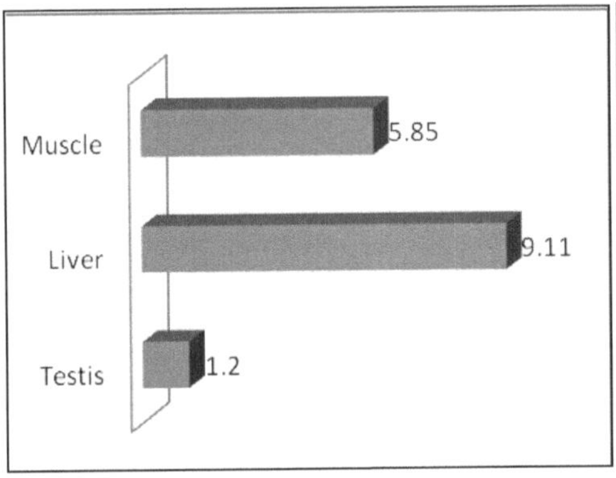

Figure 9.4: Variations of Proteins in January (Tunga Reservoir).

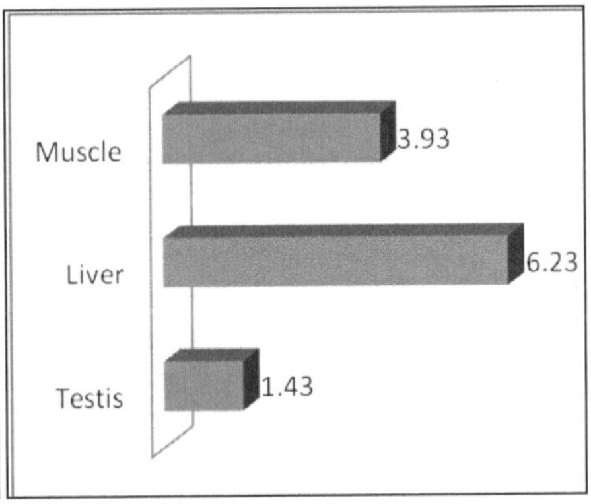

Figure 9.5: Variations of Proteins in February (Tunga Reservoir).

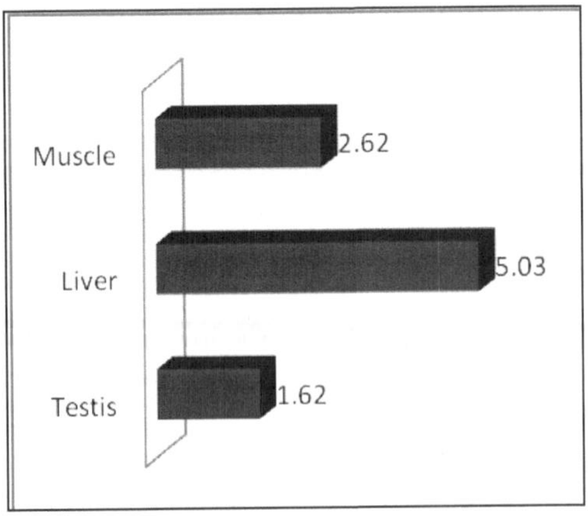

Figure 9.6: Variations of Proteins in March (Tunga Reservoir).

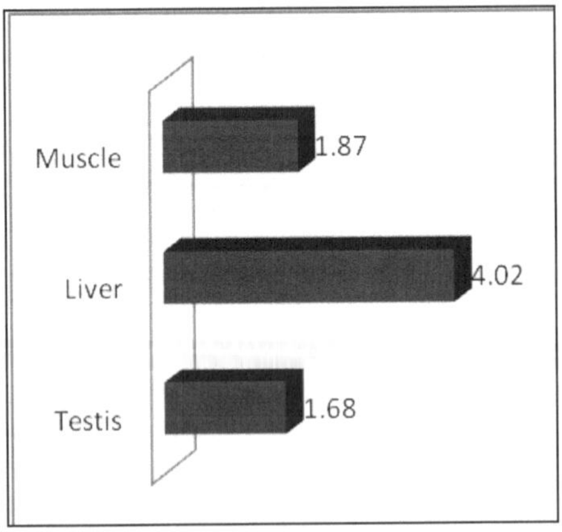

Figure 9.7: Variations of Proteins in April (Tunga Reservoir)

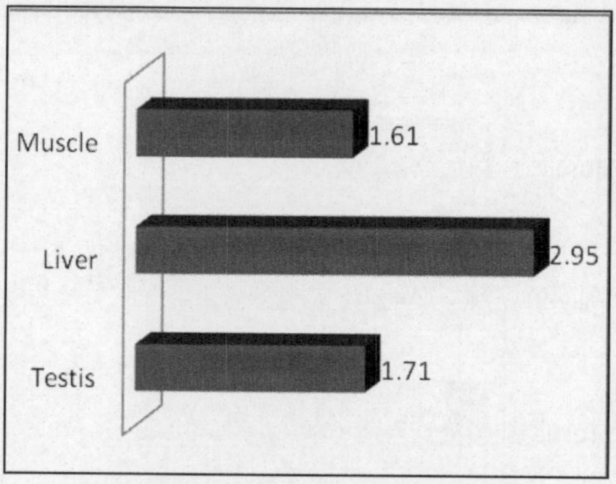

Figure 9.8: Variations of Proteins in May (Tunga Reservoir).

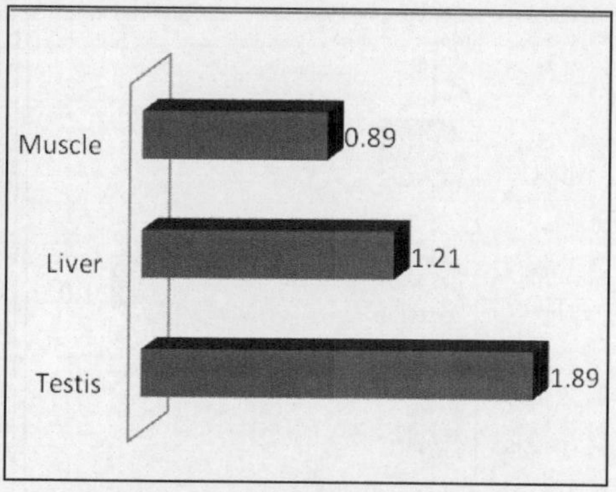

Figure 9.9: Variations of Proteins in June (Tunga Reservoir).

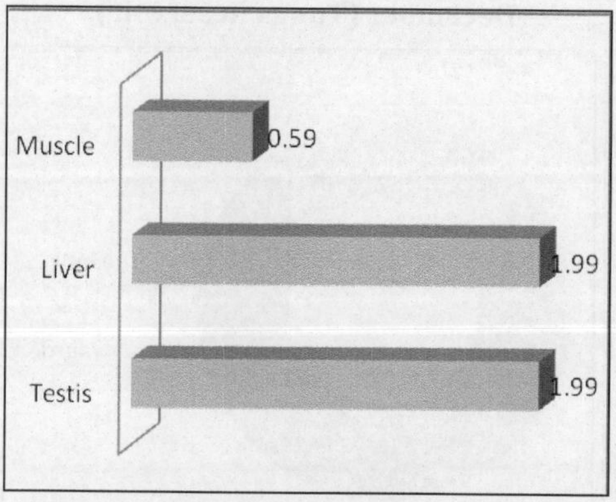

Figure 9.10: Variations of Proteins in July (Tunga Reservoir).

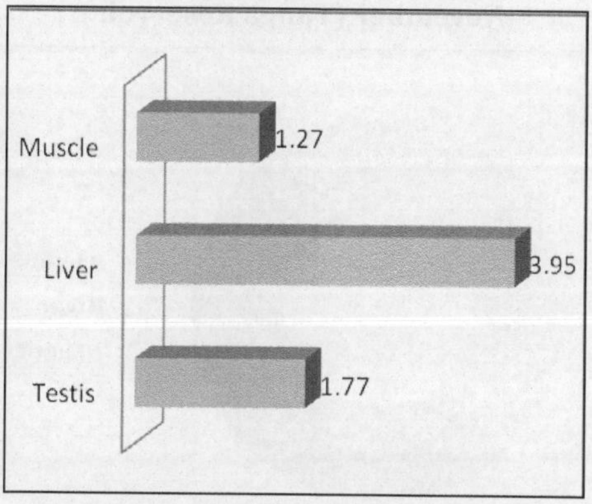

Figure 9.11: Variations of Proteins in August (Tunga Reservoir).

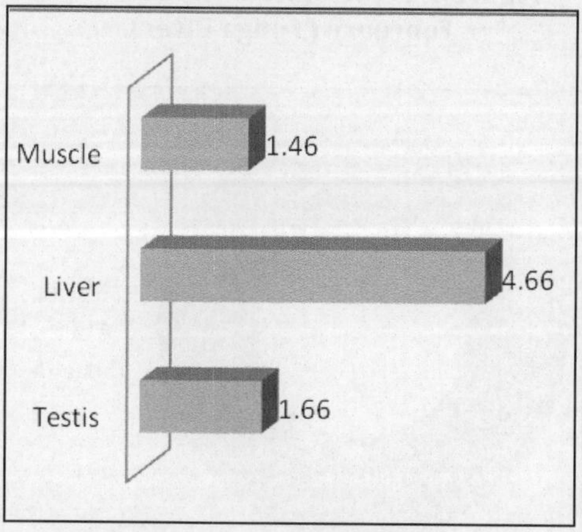

Figure 9.12: Variations of Proteins in September (Tunga Reservoir).

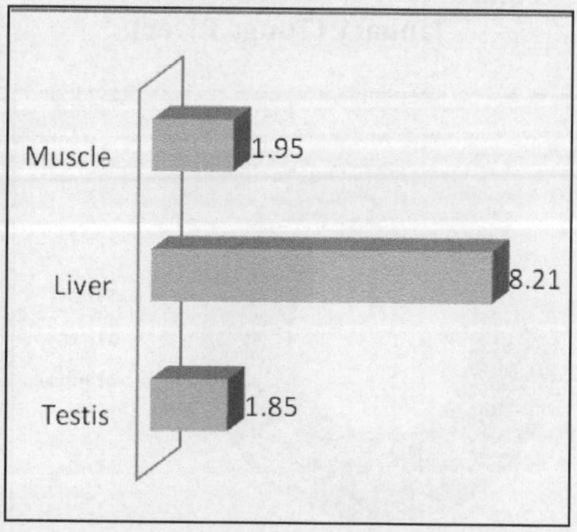

Figure 9.13: Variations of Proteins in October (Tunga Reservoir).

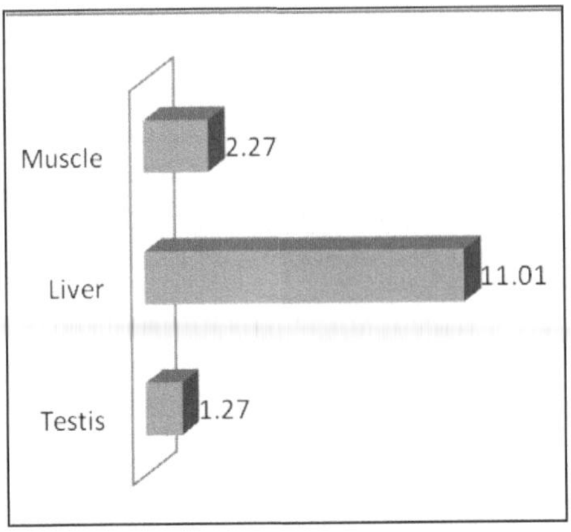

Figure 9.14: Variations of Proteins in November (Tunga Reservoir).

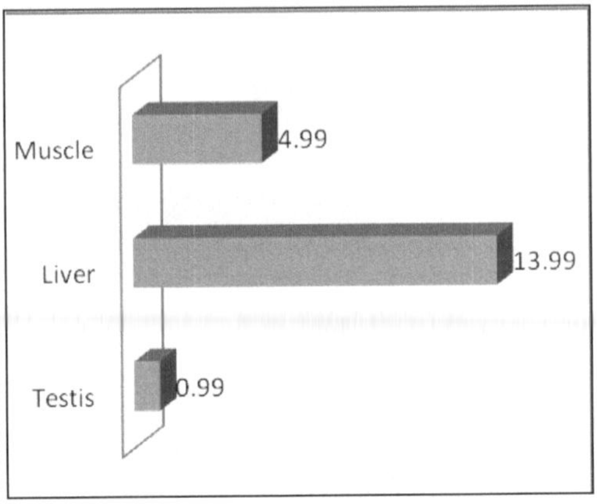

Figure 9.15: Variations of Proteins in December (Tunga Reservoir).

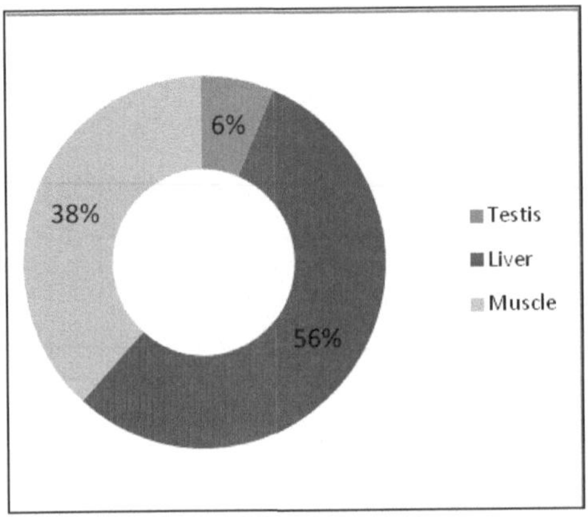

Figure 9.16: Variations of Proteins in January (Tunga River).

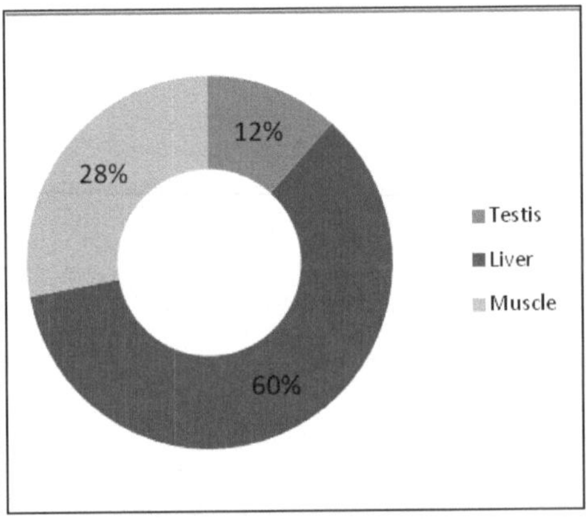

Figure 9.17: Variations of Proteins in February (Tunga River).

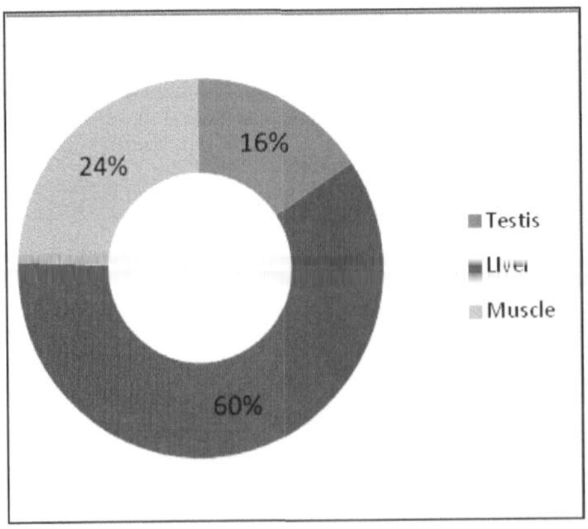

Figure 9.18: Variations of Proteins in March (Tunga River).

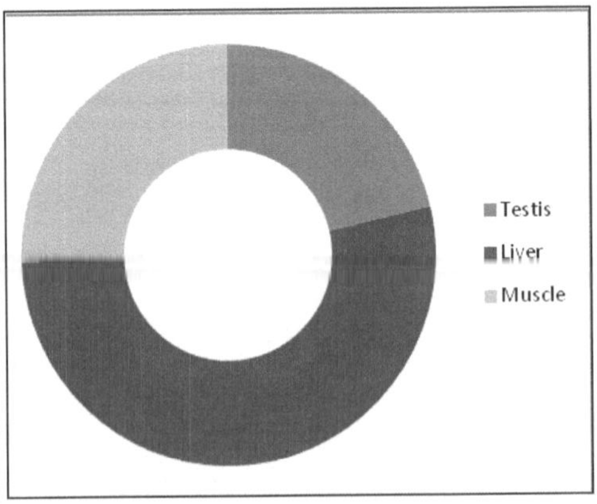

Figure 9.19: Variations of Proteins in April (Tunga River).

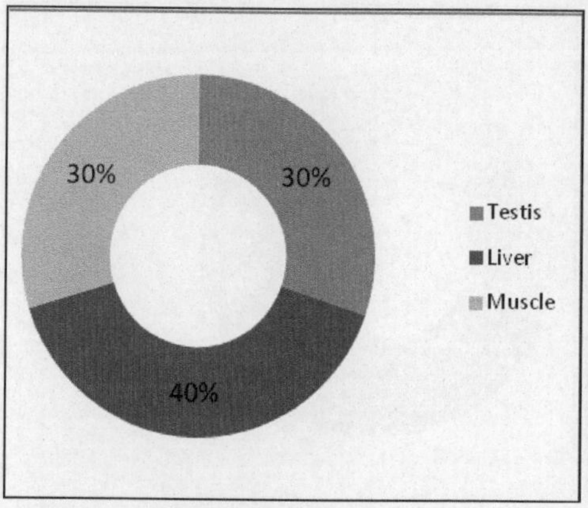

Figure 9.20: Variations of Proteins in May (Tunga River).

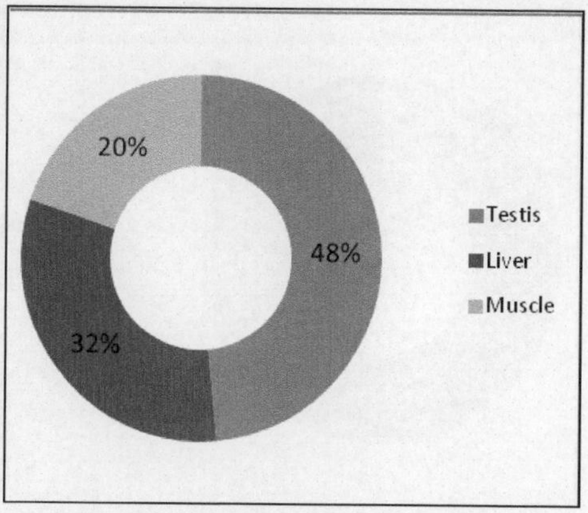

Figure 9.21: Variations of Proteins in June (Tunga River).

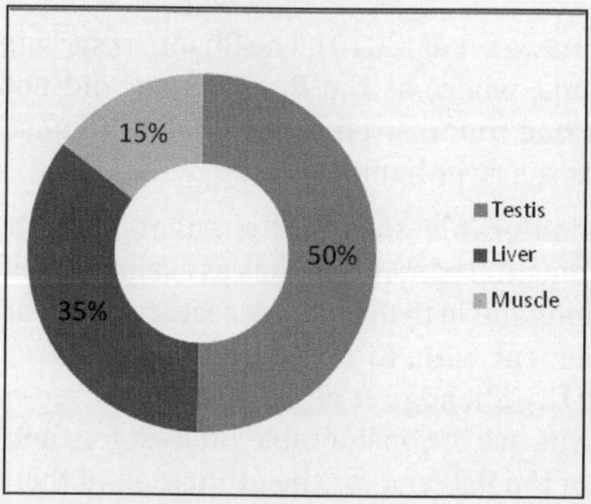

Figure 9.22: Variations of Proteins in July (Tunga River).

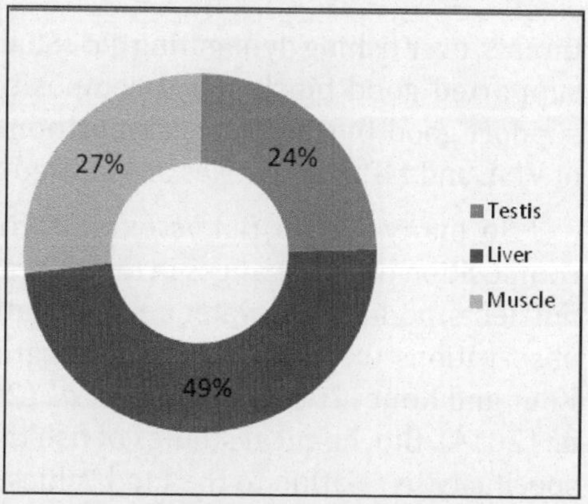

Figure 9.23: Variations of Proteins in August (Tunga River).

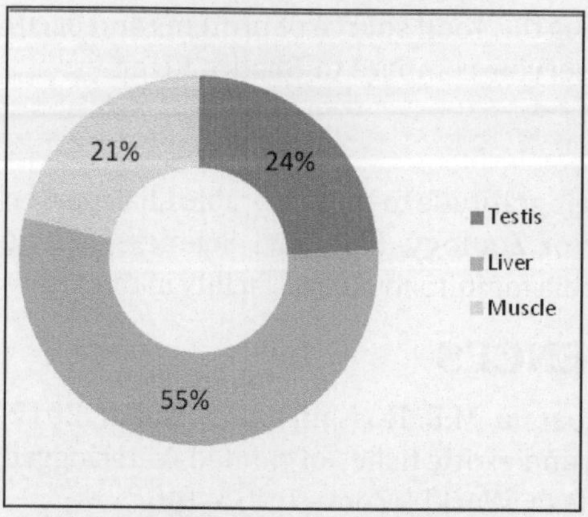

Figure 9.24: Variations of Proteins in September (Tunga River).

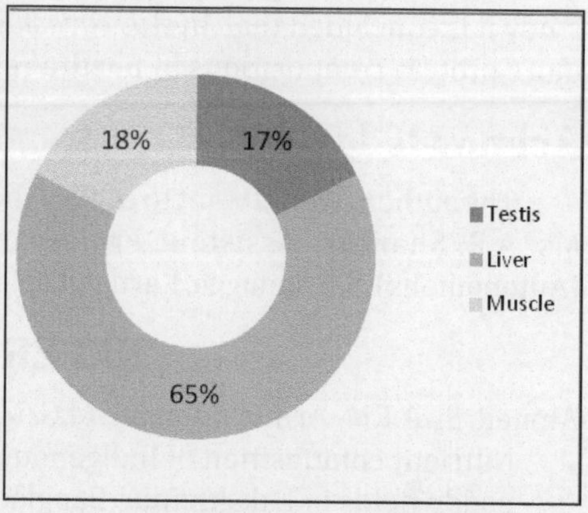

Figure 9.25: Variations of Proteins October (Tunga River).

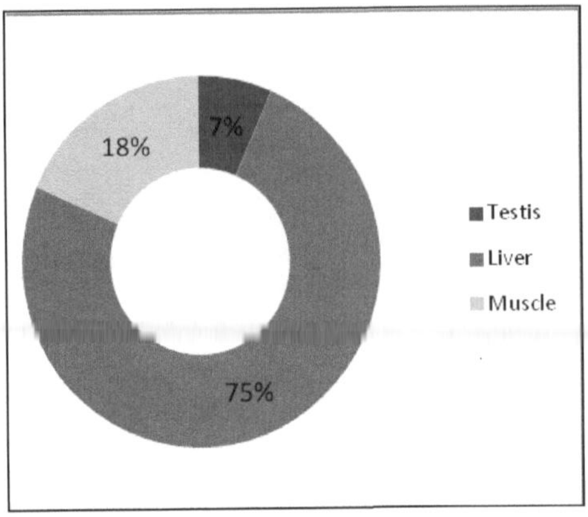

Figure 9.26: Variations of Proteins in November (Tunga River).

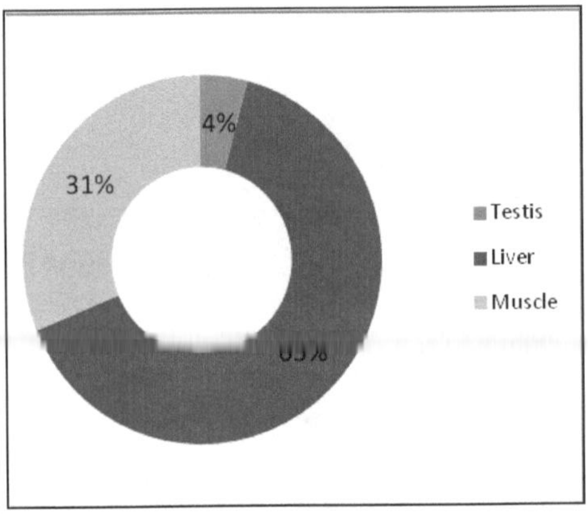

Figure 9.27: Variations of Proteins in December (Tunga River).

threats, over fishing dynamiting *etc.* (Shahnawaz *et al.,* 2010). The Bhadra reservoir supported good biochemical compositions, where as the Bhadra river did not support good biochemical compositions due to impaired water quality because of VISL and MPM industries are located at the river bank.

So many of the fishes exhibit considerable seasonal changes in their composition due to changes in the environmental conditions such as availability of nutrients, pollution of water and fluctuations and in hydrographic features, Similar observations were made here is in agreement with, Kiran and Puttaiah (2005), Kour and Kour (2006), Ulfat Jan *et al.* (2012), Shendge *et al.* (2012), Ashashree *et al.* (2014). Biochemical studies of fish tissue are of considerable interest for their specificity in relation to the food values of the fish and for the evaluation of their physiological needs at different periods of life. It is also necessary to have data about the composition of fish in order to make the best use of it as food and also to develop the technology of processing fish and fish products. Based on the current study it is concluded that *Mystus cavasius* is the good source of proteins and occur more protein compositions in Bhadra reservoir compare to Bhadra River.

Acknowledgment

The authors would want to convey their gratitude to the honorable Chairperson Ms. A.R. Shantha, Assistant Professor of Zoology, Sahyadri Science College (Autonomous), Shivamogga, Karnataka, India for infrastructural facility in this study

REFERENCES

Ahmed, S., A.F.M. Arifur Rahman, M.G. Mustafa, M.B. Hossain and N. Nahar, 2012. Nutrient composition of indigenous and exotic fishes of rainfed waterlogged paddy fields in Lakshmipur, Bangladesh. World J. Zool., 7: 135-140.

Ashashree H.M, Venkateshwarlu M and Sayeswara H.A (2013). Seasonal changes of protein in the tissues of male catfish *Mystus cavasius* (Ham) in Bhadra reservoir, Karnataka, India, IJSBPT, 4(4): 264-267.

Ashashree H.M, Venkateshwarlu M and Sayeswara H.A (2014). Seasonal changes on composition of Total lipid cntent in Muscle, Liver and testes of the male catfish *Mystus cavasius* (Ham) in Bhadra reservoir, Shivamogga, Karnataka, India, IJSR, 3(2): 20-21.

Azam K, Ali My, Asadozzaman M, Basher MZ, Hossain MM (2004), Biochemical Assessment of selected fresh fish, Journal of Biological sciences, 4 (1): 9-10.

Hemandez MD, Martinez FJ, Garcia B (2001). Sensory evaluation of farmed sharp snout sea bream (*Diplodus puntazzo*), Aquaculture international, 9: 519-529.

Kamal, D., A.N. Khan, M.A. Rahman and F. Ahamed, 2007. Biochemical composition of some small indigenous freshwater fishes from the river Mouri, Khulna, Bangladesh. Pak. J. Biol. Sci., 10: 1559-1561.

Kaur, A. and K. Kaur. (2006). Impact of nickel-chrome electroplating effluent on the protein and cholesterol contents of blood plasma of *Channa punctatus* (Bl) during different phases of the reproductive cycle. Journal of Environmental Biology, 27(2): 241-245.

Kiran (2018). Seasonal variations in biochemical composition of Salmophasia Untrachi (Day) from Bhadra Reservoir, Karnataka, International journal of Research in environmental Science, 4(3), 24-29.

Kiran, B.R. and E.T. Puttaiah. (2005). Biochemical composition of *Chela untrahi* from Bhadra reservoir, India, Geobios, 32: 93-95.

Love (1970). The chemical biology of fishes, Academic press, Inc, (London) Ltd.

Lowry, O.H, Rosebrough, N, Farr, A.L and Randall R.J (1951). Protein measurement with folin-phenol reagent. J. Biol.Chem, 193: 265-275.

Shendge, A.N, Pawar B.A and Pandarkar A.K, (2012). Variation in protein content of Freshwater catfish *Wallagoatatu* (Ham) in relation to sex and maturity cycle. J. of Experimental zoology, India vo. 15, no.1 pp 251-252.

Ulfat Jan, Mustafa Shah, Tahila Manzoor and Showkat Ahmad Ganie (2012). Variations of Protein Content in the Muscle of Fish *Schizothorax niger*, American-Eurasian Journal of Scientific Research 7 (1): 01-04, 2012.

Weatherley, A.H. and H.S. Gill, 1987. The Biology of Fish Growth. Academic Press, London, Pages: 443.

Chapter 10

Phytoplankton Diversity and Pollution Indicators of Hadadi Lake, Davangere, Karnataka

Nafeesa Begum[1] and M. Shaila[2]

Department of Botany,
Sahyadri Science College, Kuvempu University,
Shivamogga – 577 203, Karnataka
e-mail: [1]shailamresearchscholar95@gmail.com, [2]nafeesabegum30@gmail.com

Phytoplankton is microscopic plants that live in all healthy aquatic systems including freshwater ponds and lakes. Phytoplankton usually occurs in unicellular, colonial, or filamentous forms are photosynthetic. Plankton, particularly phytoplankton, has been widely used as indicators of water quality. Because of their short life span, plankton responds quickly to environmental changes. The algal diversity of Hadadi lake of Davanagere was studied. During this study 69 species of phytoplankton representing five main taxonomic groups such as Chlorophyceae, Diatoms, Desmids, Euglenoids and Blue greens were recorded. Microcystis areroginosa, Merismopedia glauca, Euglena gracils and *Scenedesmus quadricauda were the common pollution indicators. The role of phytoplankton species and their assemblage as bio-indicators reflected the pollution status of the study lake. Hence both the pollution indicator species and the beneficial species are present in the above investigated lake.*

***Keywords**: Phytoplankton diversity, Pollution indicators, Hadadi Lake, Desmids.*

Introduction

Water is one of the most important precious natural resources required essentially for the survival and health of living organisms. The term Plankton is used for assemblage of microscopic, free floating organisms in water (Suresh 2015). Nature and its contributions to a good quality of life are often perceived and valued by people in starkly different and often conflicting ways. Co-construction of assessments of the state of the world's biodiversity and the benefits it provides to humans Phytoplankton are at the base of aquatic food webs and of global importance for ecosystem functioning and services. The dynamics of these photosynthetic organisms are linked to annual fluctuations of temperature, water column mixing, resource availability and consumption. Biodiversity means the variability among all living organisms in a particular ecosystem. Ecological impacts of recent climate change on ecosystems worldwide. A major challenge in climate change research on phytoplankton succession is to understand the multiple factors, which drive ecological changes in phytoplankton communities. Climate can modify environmental factors and alter phytoplankton structure, seasonal dynamics, and taxonomic composition (Sabita Kumari *et al.*, 2016). Depends on the nature of plankton may be divided into 2 major groups namely Phytoplankton and Zooplankton.

Phytoplankton is chlorophyll bearing suspended microscopic organisms consisting mainly of Algae. The majority of the members of the phytoplankton belong to Chlorophyceae, Cyanophyceae and Bacillariophyceae group of Algae. Phytoplankton (microscopic algae) usually occurs in unicellular, colonial, or filamentous forms and is mostly photosynthetic. Plankton, particularly phytoplankton, has been widely used as indicators of water quality. Because of their short life span, plankton responds quickly to environmental changes. Some species have also been associated with noxious blooms that sometimes create offensive tasting water, odours, and/or toxic conditions. Phytoplankton encountered in bodies of water reflects the average ecological conditions of that body, and therefore, they may be used as indicators of water quality (Saha *et al.*, 2010). As well, phytoplankton are suitable organisms for the determining the impact of toxic substances on the aquatic environment because any effects on the lower levels of the food chain will also have consequences for the higher level.

Phytoplankton converts light energy to chemical energy through primary production which makes them very important in the food web. Phytoplankton plays a key role in maintaining the equilibrium between abiotic and biotic components of the wetland ecosystem (Pandey *et al.*, 2004). Use of phytoplankton density, diversity and their association as biological indicators in water quality assessment and trophic status. Species composition, abundance and diversity of phytoplankton are monitored by environmental factors like physico-chemical properties of water, meteorological properties of particular region, morphometric and hydrographic characters of the wetland ecosystem. Phytoplankton plays a role in regulating atmospheric temperature via photosynthesis. Phytoplankton plays a role in regulating atmospheric temperature via photosynthesis. Phytoplankton is the primary producers forming the first tropic level in the food chain. Diversity of planktonic organisms is quite high in fertile standing water bodies. Planktons are very sensitive to the environment they live in and any alteration in the environment leads to the change in the plankton communities in terms of tolerance, abundance, diversity and dominance in the habitat. (Aarti Devi *et al.,* 2013).

Plankton respond quickly to environmental changes because of their short life cycle, hence, their species composition and dynamics are more likely to indicate the quality of the water which they are found. The present study is aimed at evaluating the phytoplankton index of the study area as water quality criteria with reference to freshwater bodies polluted by various anthropogenic activities.

In aquatic ecosystems phytoplankton are the principal primary producer's forms the first trophic level of the aquatic food chain (Nasser and Sureshkumar, 2014). Hence they play an important role in the food webs as they provide food for zooplankton and other aquatic fauna (Jagadeeshappa and Kumara, 2013). Their presence is also essential to support a healthy aquatic ecosystem, and is frequently used as ecological indicator for the ecological health and the stress effects of chemical contaminants in aquatic ecosystems. Phytoplankton diversity and dynamics in open lakes is largely dependent on abiotic factors such as nutrient availability, water temperature, light conditions and transparency, along with biotic interactions such as predation and competition. Plankton respond quickly to environmental changes because of their short life cycle, hence, their species composition and dynamics are more likely to indicate the quality of the water which they are found. In this paper an attempt has been made to study the phytoplankton diversity and Pollution indicators in Hadadi Lake.

Materials and Methods

Study Area

Hadadi Lake is situated 14kms away from Davangere city. It lies between 14°22' N L and 75° 50' E L. Total water spread area of this water body is 139 hectares. Rain fall is the major source of water and Bhadra right bank channel is the other source. The water body is mainly used for growing agricultural crops. Aquaculture is the other activity in this water body.

Phytoplankton Collection and Analysis

A water sample measuring one litre was collected for studying planktonic composition. Water sample was fixed with 20 ml of 1 per cent Lugol's iodine solution and kept 24 hours for sedimentation. 100mL of sample is subjected to centrifuge at 1500 rpm for 20 minutes and concentrate was made up to 20 ml with 5 per cent formalin used for further investigation. Identification of phytoplankton up to species level was done by referring the standard monographs of Desikachary (1959), Prescott (1982) Seatt and Prescott (1961), Philipose (1960), Gandhi (1998) Taylor *et al.* (2007).

Results and Discussion

The results of the indices for the Hadadi Lake are given in Table 10.1. The distribution of species indicates that Hadadi Lake has the 69 Species of phytoplankton. The algae were expressed as organisms per ml for the purpose of calculating diversity indices. A total of 20 algal species belonging to Cholrophyceae, 20 species from Bacillariophyceae, 8 Species from Desmids, 10 Species from Euglenoids, 11 species from Blue greens were recorded as shown in Table 10.1. Diatom cells often remained for a long duration and hence can serve as an important indicator of organic pollution as well as anthropogenic activity (Hosmani, 2010).

Table 10.1: Phytoplanktons Recorded in Hadadi Lake

Sl.No.	Phytoplanktons	Sl.No.	Phytoplanktons
	Chlorophyceae	6	*Fragillaria*
1	*Actinastrum sp.*	7	*Gyrosigma accuminatum*
2	*Arthrodesmus sp.*	8	*Gyrosigma gracilis*
3	*Chlorella vulgaris*	9	*Gyrosigma elongate*
4	*Coelastrum sp.*	10	*Melosira granulata*
5	*Crucigenia crucifera*	11	*Navicula pupula*
6	*Crucigenia retangularis*	12	*Navicula cuspidate*
7	*Chlamydomonassp*	13	*Navicula pigmea*
8	*Eudorinaelegans*	14	*Navicula cryptocephala*
9	*Pediastrum simplex*	15	*Navicula menuisculus*
10	*Pediastrum ovatum*	16	*Navicula radiosa*
11	*Pediastrum constrictum*	17	*Navicula dicephala*
12	*Scenedesmus dimorphous*	18	*Pinnularia microstauron*
13	*Scenedesmus oblique*	19	*Surirella capronii*
14	*Scenedesmus protuberans*	20	*Cyclotella stelligera*
15	*Scenedesmus plastydiscus*		**Desmids**
16	*Scenedesmus quadricauda*	1	*Cosmarium retusiformi*
17	*Scenedes mussmithi*	2	*Cosmarium capitulum*
18	*Scenedes musgracil*	3	*Cosmarium constractum*
19	*Selenastrum westi*	4	*Cosmarium depressum*
20	*Tetra daedrongracile*	5	*Cosmarium nudum*
	Bacillariophyceae	6	*Cosmarium panctualatum*
1	*Acanthes* sp.	7	*Micrasterias* sp.
2	*Anomoeonies sphaeophora*	8	*Staurastrum* sp.
3	*Closterium* sp.		**Euglenophyceae**
4	*Cymbella cistula*	1	*Euglena acus*
5	*Cyclotella stelligera*	2	*Euglena obtus*

Sl.No.	Phytoplanktons	Sl.No.	Phytoplanktons
3	*Euglena spirogyra*	2	*Anacystis* sp.
4	*Euglena gracil*	3	*Anabaena aphnizomenoides*
5	*Euglena accutissima*	4	*Anabaena spiroides*
6	*Phacus orbicularis*	5	*Aphanocapsa* sp.
7	*Phacus meson*	6	*Chroococcus turgidus*
8	*Strombomonas gibberosa*	7	*Merismopedia glauca*
9	*Strombomonas* sp.	8	*Microcystis viridis*
10	*Trachelomonas robasta*	9	*Nostoc microscopium*
	Cyanophyceae	10	*Oscillatoria tennuis*
1	*Agmenellum* sp.	11	*Phormidium* sp.

In the present study the species of Chlorophyceae and Bacillariophyceae were dominating represented when compared to other groups of the study as shown in Figure 10.1, the occurrence of Euglenophyceae and Dinophyceae were inversely related to Chlorophyceae. Chlorophyceae was the most dominating group.

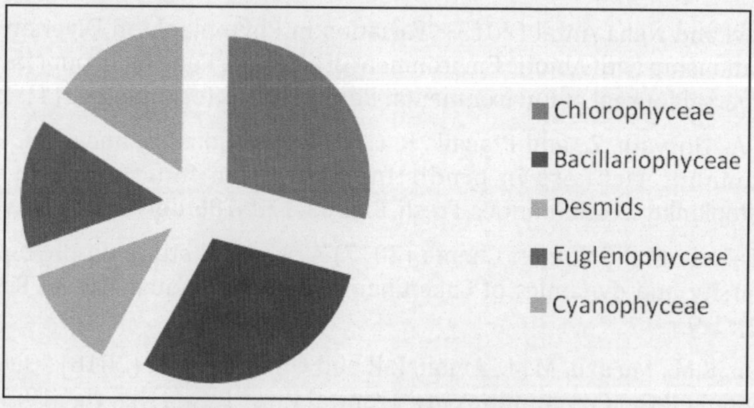

Figure 10.1: Phytoplankton Diversity.

Temperature plays an important role in the periodicity of blue green algae as emphasized by Mahar *et al.* (2004). Chlorococcales occur as greenish scum on the surface of stagnant water. Factors like high temperature and bright sunlight are favourable for the growth of chlorococcales. Among phytoplankton community, diatoms play a very important role ecologically as they comprise of major components of producers in wetland ecosystem. Diatoms are ubiquitous, unicellular microorganisms form the basic bulk of planktonic population in freshwaters characterised by siliceous cell wall (Round *et al.*, 1990). Sabata (2004) recorded highest number of diatoms during summer coupled with silica. Bluegreens are cosmopolitan in nature play a significant role in wetland ecosystem. Blue-greens exhibited heavy growth in polluted water bodies and dominated over Chlorophyceae and Bacillariophyceae. Pollution Tolerant species like *Scenedesmus quadricauda, Tetraedonsps, Microcystis aeroginosa, Phacus* sp., *Euglena* sp. and *Coelastrum* sp. were observed (Figure 10.1). These species are used as the bestindicator of pollution and are associated with highest degree of civic pollution.

Conclusion

Diversity measures are more useful in Lake Ecosystem, which harbour a large variety of algal species in general and species diversity within genera. The role of phytoplankton species and their assemblage as bio-indicators reflected the pollution status of the study lake. *Microcystis areroginosa, Merismopedia glauca, Euglena gracils* and *Scenedesmus quadricauda* were the common pollution indicators. Hence both the pollution indicator species and the beneficial species are present in the above investigated lake. Hence environmental monitoring makes use of the fact that polluted or stressed communities are characterized by a change in the species abundance. The present study concludes that the maximum diversity of phytoplankton's present in the Hadadi Lake of Davangere District.

REFERENCES

Aarti Devi and Neha Antal (2013) "Variation in Phytoplankton Diversity and Its Relationship with Abiotic Environment of a Temple Pond in Birpur (J&K) India" Universal Journal of Environmental Research and Technology, 3(1): 181-189.

Abonyi, A., Horváth, Z., and Ptacnik, R. (2018). Functional richness outperforms taxonomic richness in predicting ecosystem functioning in natural phytoplankton communities. Fresh. Biol. 63, 178–186. doi: 10.1111/fwb.13051

AddisuFekadu and Solomon Chanie (2017)"A seasonal study on phytoplankton diversity and dynamics of Lake Chamo, Ethiopia", Aquat. Living Resour. 30 (40): 1-9.

Adelakun, K.M., Mu'azu, M.M., Amali, R.P. and Omotayo, O.L. (2016) "Diversity of Phytoplankton Communities in a Tropical River Basin, Nigeria", 19, 52-56.

Bode, A., Alvarez, M., Ruíz-Villarreal, M., and Varela, M. (2018). Time Series of Hydrographic, Chemical and Primary Production Variables for a Shelf Station off A Coruña (NW Spain): 1989–2016. PANGEA. Available online at: https: // doi.pangaea.de/10.1594/PANGAEA.885413.

Herni Effendi, MujizatKawaroe, DeaFauziaLestari, Mursalina, Tri Permadia, 2016 "Distribution of phytoplankton diversity and abundance in Mahakam Delta, East Kalimantan" Procedia Environmental Sciences (Elesvier), 33, 496 – 504.

Jaime Otero, Xosé Antón Álvarez-Salgado and Antonio Bode (2020) Phytoplankton Diversity Effect on Ecosystem Functioning in a Coastal Upwelling System, Frontiers in Marine Science, 7: 1-15.

Mahadev J, G. Sugeetha, N.S. Pankaja and K.V. Shivakumar (2019) "Alga as an Indicator of Water Quality in Freshwater Bodies of Mysore, Karnataka, India", Int. J. Curr. Microbiol. App. Sci, 8(3): 2264-2271.

Malik D.S. and Umesh Bharti (2012) "Status of plankton diversity and biological productivity of Sahastradhara stream at Uttarakhand, India" Journal of Applied and Natural Science 4 (1): 96-103.

Nafeesa Begum, J. Narayana and H.A. and Sayeswara A (2010) "Seasonal Study Phytoplankton diversity and pollution indicators of Bathi pond near Davanagere City, Karnataka (India)", Environment Conservation Journal 11(3) 75-80.

Nasser KMM, Sureshkumar S. (2014). "Seasonal variation and biodiversity of Phytoplankton in Parambikulam Reservoir, Western Ghats, Kerala", Int J pure Appl Bio Sci 2: 272–280.

Rajkumar M, P. Perumal, V. Ashok Prabu, N. Vengadesh Perumal and K. Thillai Rajasekar, (2009), "Phytoplankton diversity in Pichavaram mangrove waters from south-east coast of India", Journal of Environmental Biology, 30(4): 489-498.

SabitaKumari P, Gayathri S and Ramachandra Mohan M, (2018) "Phytoplankton Diversity in Bangalore Lakes, Importance of Climate Change and Nature's Benefits to People", Journal of Ecology and Natural Resources, 2(1): 118-125.

SahaS.B.; Bhattacharya, S. B.; Choudhury, A. (2000). Diversity of phytoplankton of sewage pollution brackish water tidal ecosystems. Environ. Biol. 21 (1): 9-14.

Shivakumara L.V and Pramod V. Pattar (2015) "Diversity of Phytoplanktons in Rice Fields of Davangere Taluk, Karnataka", Journal of Marine Science: Research and Development, 5 (3): 1-4.

Sudeep B M, S. Srikantaswamy and Shankar P. Hosmani (2008) "The Study of Phytoplankton Dynamicsin Two Lakes of Mysore, Karnataka State, India, Nature Environment and Pollution Technology, 7(4), 697-702.

Suresh (2015) "Multiplicity of phytoplankton diversity in Tungabhadra River near Harihar, Karnataka (India)", Int.J.Curr.Microbiol.App.Sci, 4(2): 1077-1085.

Chapter 11

Distribution of Octochaetona Species in the Hyderabad-Karnataka Region of Karnataka, India

Heena Mubeen[1] *and Shankerappa S. Hatti*[2]

[1]*Department of Zoology,*
Gulbarga University, Kalaburagi – 585 106, Karnataka
[2]*Department of UG and PG Studies and Research in Zoology,*
Government College (Autonomous), Kalaburagi – 585 105, Karnataka
e-mail: [1]*heenamubeen323@gmail.com;* [2]*hattishankerappa@gmail.com*

A random field survey was conducted in the selected habitats like near the gutter, near bore well, garden, cultivated land and drainage system across 6 districts of H-K region. In total 8 Octochaetona species are identified from 23 different sites. They are Octochaetona compta, Octochaetona paliensis, Octochaetona prashadi, Octochaetona suriensis, Octochaetona parva, Octochaetona beatrix, Octochaetona albida and Octochaetona phillotti. Among these O. compta, O. philloti and O. suriensis are reported for the first time reported from Karnataka state. All the Octochaetona species recorded in our study area during our study period are endogeic and indigenous to India. Soil chemical parameters of their habitat were in the following range, pH 7.55±0.21 to 7.9±0.89, Electrical Conductivity (EC) 0.26±0.03 to 0.80±0.65mohs/cm, Organic Carbon (OC) 1.12±0.40 to 3.325±0.26 per cent, Nitrogen (N) 37.52±13.27 to 111.39±8.77Kg/acre, Phosphorous (P) 17.4±5.89 to 25.59±14.66Kg/acre and Potassium (K) 187±50.41 to 296±82.02 Kg/acre.

***Keywords**: Diversity, Earthworm, Habitat, Soil.*

Introduction

Octochaetona belong to the phylum Annelida, class Oligochaeta, order Haplotaxida and family Octochaetidae. In recent years, the diversity of Indian earthworms has been mainly studied by Julka (1988). He revised the Indian members of the family Octochaetidae in the publication "Fauna of India" providing illustrated descriptions of 154 taxa including 6 new genera and 16 new species. Totally 16 different species of the genus *Octochaetona* have been reported in India by Begum and Ismail (2004), Kathireswari *et al.* ([illegible]), G. Tripathi and P. Bharadwaj (2004), J.M. Julka (1995), Om Prakash (2017), Palanisamy Kathireswari, Jatinder M. Julka and John W. Reynolds (2005), Rahul Paliwal (2008) Madhab C. Dash and K.G. Saxena, (2012), Mohan V.C., Watts P. and Kaur (2013) and Sharma R. K and Bharadwaj Poonam (2014).

Presence of *Octochaetona* species in Karnataka was first reported by Radha D Kale and R.V. Krishnamoorthy (1978). Subsequently more information regarding *Octochaetona* species was added by Bano and Kale (1991), Kale and Karmegam (2010), Siddaraju *et al.* (2010), Muddaiah *et al.* (2013). Very little information is available about the earthworms from The H-K region. The first attempt was made by Shankerappa S. Hatti (2013) to study the earthworm diversity of Gulbarga city and he reported 4 earthworm species in his studies. 14 earthworm species were reported from Koppal district by Mubeen and Hatti (2018). So far the occurrence of four *Octochaetona* species *i.e., Octochaetona albida, Octochaetona beatrix, Octochaetona paliensis* and *Octochaetona parva* have been reported from southern part of Karnataka state. However present diversity and distribution status any of these species is not yet thoroughly studied in Hyderabad- Karnataka region.

Many researchers have confirmed carried out studies on earthworms in various parts of the world mentioning soil diversity is important in order to sustain the ecosystem processes. (Ismail and Murthy, 1985; Tsai *et al.,* 1999, 2000; Blakemore, 2002; Chang and Chen, 2004, 2005; Blakemore *et al.,* 2006; Sautter *et al.,* 2006; Julka *et al.,* 2009; Yogendra and Keshav Singh, 2013; Makin *et al.,* 2014; Sharma and Poonam, 2014; V. Mariappan *et al.,* 2013; Rinku Goswami and C.K. Mondal, 2015; Deepthi and Kathireshwari, 2016; Mubeen and Hatti, 2018)

Presently there is great paucity in the information of *Octochetona* species of H-K region. With the intention to find out locally available *Octochetona* species, a survey was conducted in H-K region.

Materials and Methods

Study Area

The H-K region is the second largest arid region after parts of Rajasthan in India. It is located at 17°35' and 18° 25' North latitudes and75° 40' and 77° 11' East longitude in the north-eastern part of the Karnataka state. It comprises of Bidar, Kalaburagi, Yadgir, Raichur, Koppal and Bellary districts. Entire Bidar district is formed of solidified lava and forms the part of Deccan trap of H-K Region. Black soil is predominant in this region. Apart from Black soil red soil,mixed red and black soils, medium black soils, deep black soils and red sandy soils are the major type of soils are the types of soils found in this region. It experiences semi-arid climate with extreme summer. The temperature varies in this region between 15 °C and 41°C with an average rainfall of 692 mm. Very hot summers are the key causes for less fertile output in terms of agriculture in spite of two major rivers Krishna and Tungabhadra, flowing through the region. Mullameri, Bennithora, Kagina and Bhima are tributaries to Krishna river flows through Bidar, Kalaburagi and Yadgir districts. Manjra is a tributary to Godavari River which flows through Bidar district.

Earthworm and Soil Sampling

Random earthworm sampling was carried out once from July 2017-October 2017 during rainy season. The collections were made by digging and hand sorting method (Julka 1988) from the selected habitats like gutter, near bore well, garden, drainage system and cultivated land of different district of the H-K region. Earthworm samples were collected, each from 25x25cm and 20cm deep. Physical parameters like temperature, humidity and location were recorded. Morphological observations like body colour, length and width of the collected earthworm were also noted down. The collected earthworms were anaesthetized with 70 per cent ethyl alcohol and later fixed by stretching in a cotton tray containing 5-10 per cent formalin, left for overnight and preserved in 5-10 per cent formalin (Julka and Paliwal 1993). The collected samples were divided into two groups, one group was taken to the laboratory for identification and the other was reconfirmed by Oligochaetologist Rahul Paliwal, (Retd.) scientist, Zoological Survey of India (Kolkata).

In the Laboratory

Color, length and width of the preserved earthworms were once again recorded. Prior to dissection morphological characters like size, prostomium, dorsal pores, clitellum, setal arrangement, genital pores, and genital markings were noted down using stereozoom microscope. Earthworm was then pinned dorsally to the dissecting tray using small ball headed dissecting pins by taking care to avoid damage to the prostomium. The photographs of the spermathecal pore region and the genital region of the earthworms were taken using digital camera fixed to stereo zoom microscope. A longitudinal slit was made dorsally slightly away from the median line using a scalpel to avoid damage to the dorsal blood vessels. Anatomical characters such as gizzard septa, calciferous glands, beginning of intestine, position and type of typhlosole were noted down. Earthworms were Identified using taxonomic key charts (Stephenson 1923; Gates 1972; Julka 1988).

Soil samples of their habitats were collected for its chemical analysis. Soil pH was analyzed by Potentiometry (Piper 1966), Electrical Conductivity (EC) by Solubridge conductivity meter method (Black and Evans 1965), Organic Carbon

(OC) by Walkley and Black's wet oxidation method (Jackson 1967), Nitrogen (N) by Alkaline potassium permanganate method (Subbaiah and Asija 1956), Phosphorous (P) by Olsen's method (Jackson 1967) for neutral to alkaline soils and by Bray's method (Bray and Kurtz 1945) for acidic soils and Potassium (K) by Flame photometer (Jackson 1967). Soil analysis was carrie1d out with the help of Agriculture Soil testing laboratory of Kalaburagi

Results

Our results show 8 species of the *Octochaetona* are identified from 23 sites of the H-K region (Figure 11.1). They are *Octochaetona compta* (Gates, 1962), *Octochaetona paliensis* Stephenson, 1920, *Octochaetona prashadi* Gates, 1962, *Octochaetona suriensis* Michaelsen, 1910, *Octochaetona parva* Gates, 1962, *Octochaetona beatrix* Beddard, 1902, *Octochaetona albida* Gates, 1962and *Octochaetona phillotti* Michaelsen, 1907.

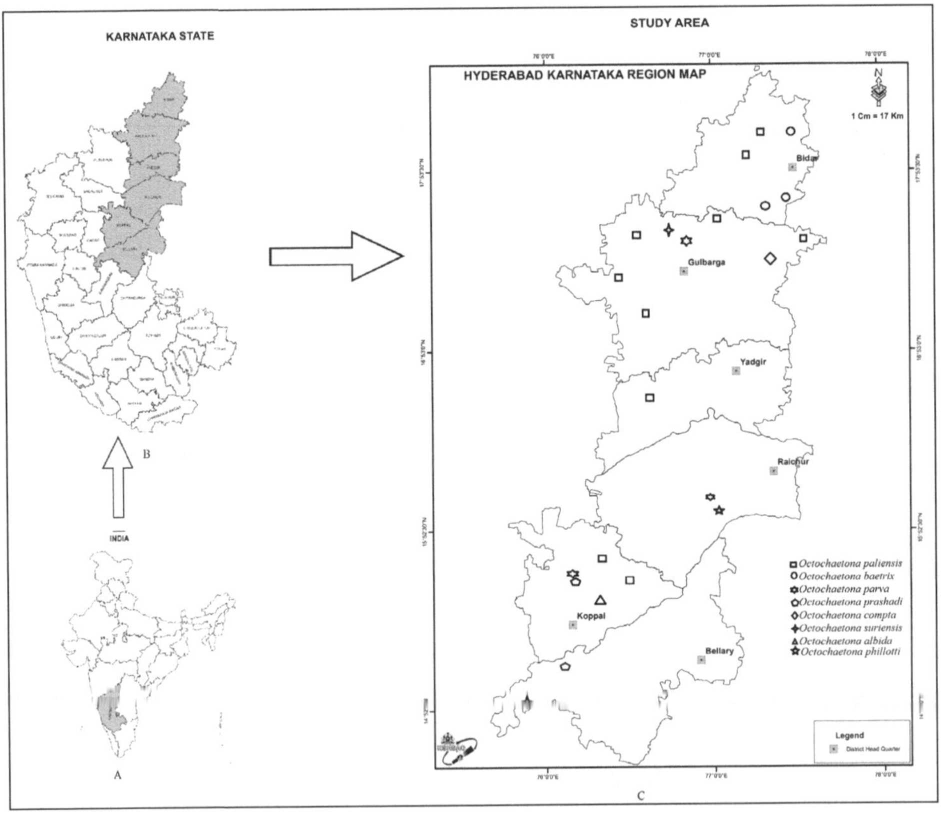

Figure 11.1: (A) Map of India; (B) Map of Karnataka State; (C) Map of Hyderabad-Karnataka Region Showing the Diversity and Distribution of Species of *Octochaetona*.

Genus *Octochaetona* Gates, 1962

Diagnosis

Setal arrangement lumbricine. Male pores paired in the seminal grooves on XVIII; paired prostatic pores at the ends of seminal grooves on XVII and XIX. Oesophagus with a single gizzard in VI or in a space between septa IV/V and VII/VIII or VIII/IX or IX/X. One pair of discrete extramural asymmetrical calciferous glands opening into the gut close to the attachment of septum XV/XVI; supra-intestinal glands absent; typhlosole lamelliform and bifid ventrally. Astomate micromeronephridia with paired enteronephric tufts in IV; several are exonephric on the body wall in v and in the posterior segments. Slightly enlarged and stomate micromeronephridia with preseptal and intrasegmental funnels present in the caudal segments. Megameronephridia absent. Table 11.1 depicts the major morphological and anatomical differences between *Octochaetona* species recorded in our study area.

Octochaetona paliensis (Figure 11.2) differs from the other *Octochaetona* species in having body length 35-92mm, body diameter 2-3.5mm, 119-182 body segments. Proepilobic prostomium. Copulatory setae surrounded by tumescences. Pineal setae ornamented with circles of spines, tip bluntly pointed. Last pair of heart in 12.

Distribution: Karnataka; Madhya Pradesh, Hyderabad, Vijaypuri, Andhra Pradesh, Rajasthan and Uttar Pradesh.

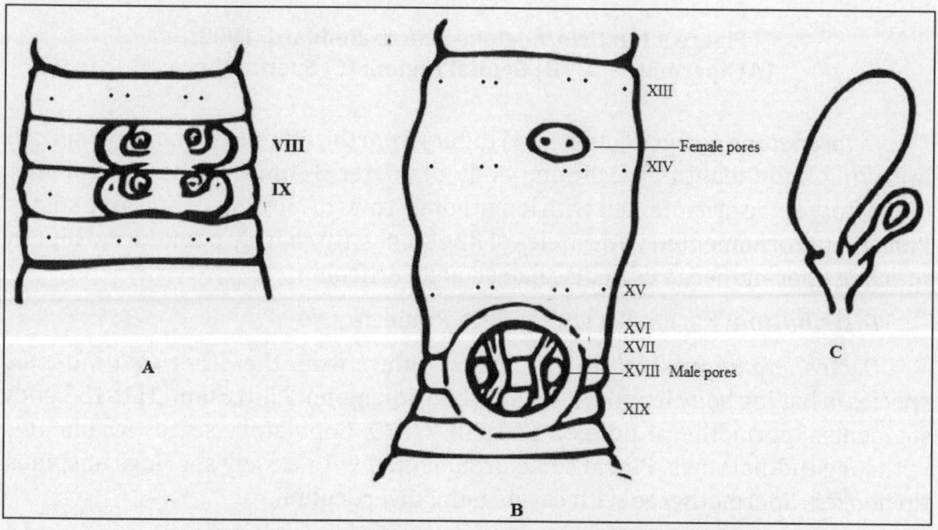

Figure 11.2: *Octochaetona paliensis* Gates, 1962: (A) Spermathecal pore region; (B) Genital region; (C) Spermatheca.

Octochaetona beatrix (Figure 11.3) differs from the other *Octochaetona* species in having body length 40-134 mm, body diameter 2-5mm, 133-197 body segments. Copulatory setae ornamented with few scattered sigmoid excavations. Pineal setae ornamented with sparse triangular teeth tip. Each spermathecae with a spheroidal, shortly pyriform, oval or flattened and shelf-like, shortly stalked ental diverticulum. Metandric testis. 1 pair of Seminal vesicles in 12.

Distribution: Karnataka, Uttar Pradesh, Madhya Pradesh, Haryana, Punjab, Himalaya and Western Ghats region of India

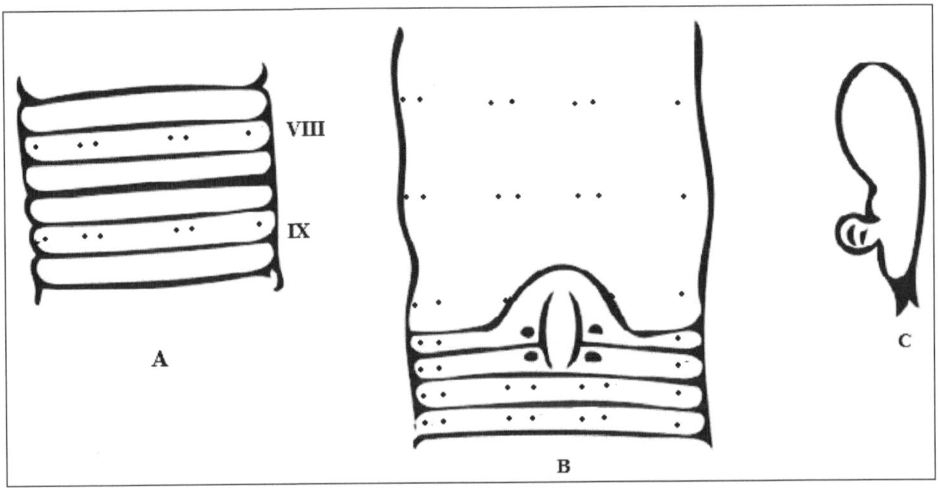

Figure 11.3: *Octochaetona beatrix* Beddard, 1902: (A) Spermatheca; (B) Genital region; (C) Spermatheca.

Octochaetona parva (Figure 11.4) differs from the other *Octochaetona* species in having body length 70-100mm, body diameter 3-4mm, 80 body segments. Copulatory setae ornamented with longitudinal rows of fairly stout and long spines. Pineal setae ornamented with circles of fine spines, tip sharply pointed and bluntly rounded Spermathecae with disc like ectal diverticulum.

Distribution: Karnataka and Andhra Pradesh.

Octochaetona prashadi (Figure 11.5) differs from the other *Octochaetona* species in having body length 51-61mm, body diameter 2.5-3.5mm, 150-155 body segments. Spermathecal pores 2 pairs in 7/8/9. Copulatory setae Ornamented with longitudinal rows. Pineal setae ornamented with closely set rings of spines, tip hooked. Spermathecae with cavate ental diverticulum.

Distribution: Maharashtra.

Octochaetona suriensis (Figure 11.6) differs from the other *Octochaetona* species in having body length 60-140mm, body diameter 2.5-6mm, 111-180 body segments. Copulatory setae ormamented with longitudinal rows of spines, tip claw

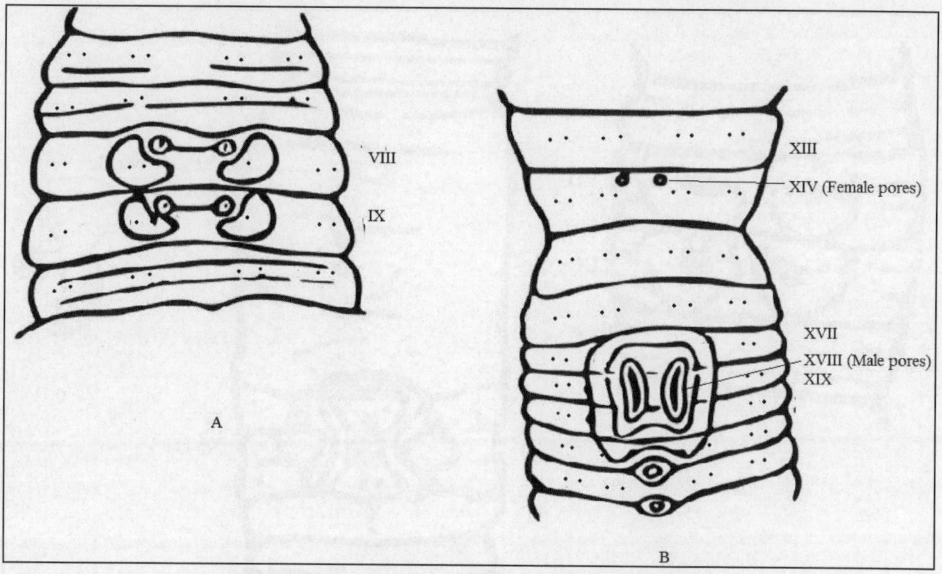

Figure 11.4: *Octochaetona parva* Gates, 1962: (A) Spermathecal pore region; (B) Genital region.

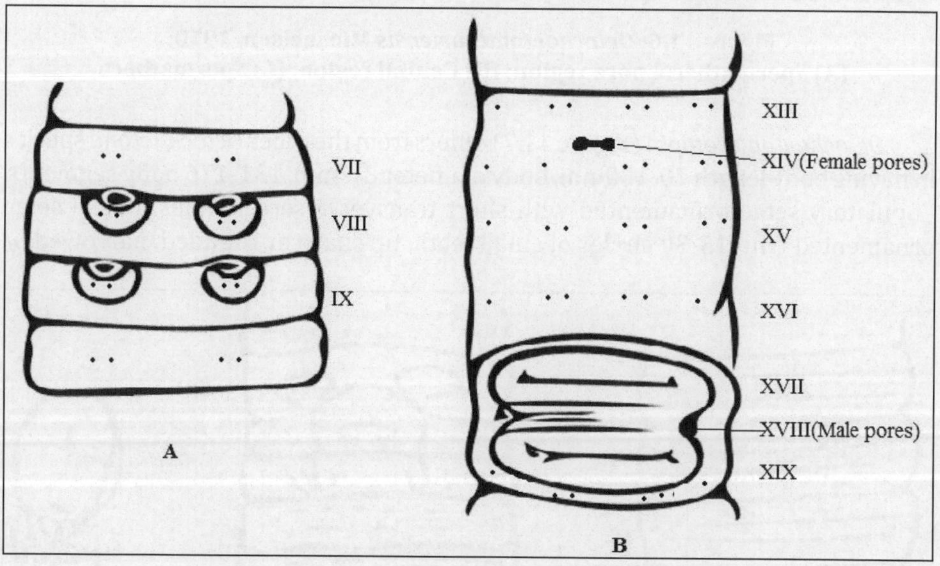

Figure 11.5: *Octochaetona prashadi* Gates, 1962: (A) Spermathecal pore region; (B) Genital region.

shaped. Copulatory setae ornamented with few longitudinal rows of triangular teeth, tip pointed or claw shaped. Each spermathecae with shortly stalked, multiloculate ental diverticulum, duct longer than ampulla.

Distribution: Tamil Nadu, Madhya Pradesh, Himalaya and Western Ghats region of India.

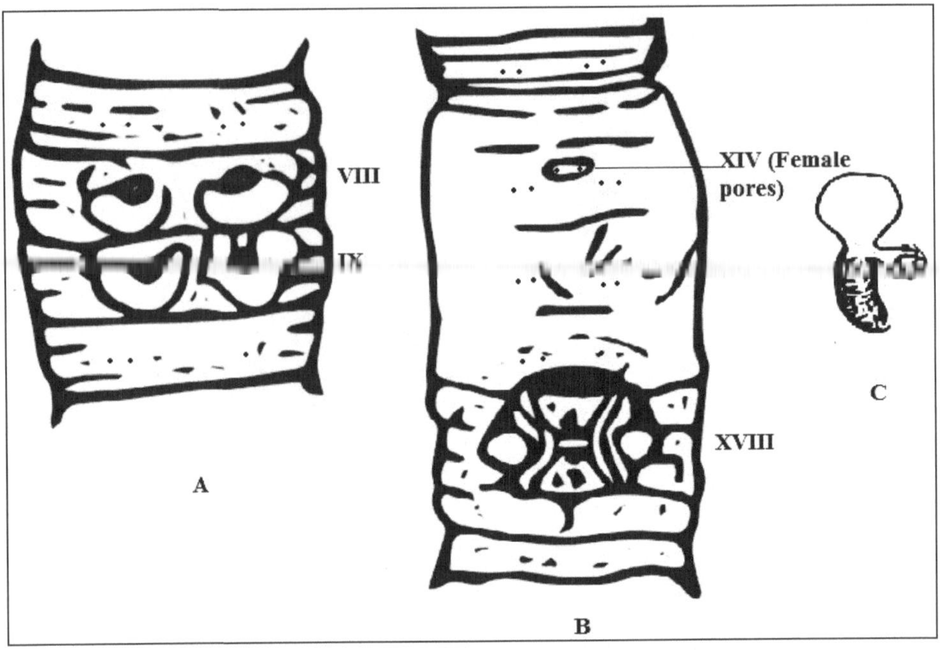

Figure 11.6: *Octochaetona suriensis* Michaelsen, 1910: (A) Spermathecal pore region; (B) Genital region; (C) Spermatheca.

Octochaetona compta (Figure 11.7) differs from the other *Octochaetona* species in having body length 70-150mm, body diameter 3-7mm, 131-176 body segments. Copulatory setae ormamented with short transverse serrate lines. Pineal setae ornamented with 15-30 circles of small teeth, tip sharp or rounded, narrowed or

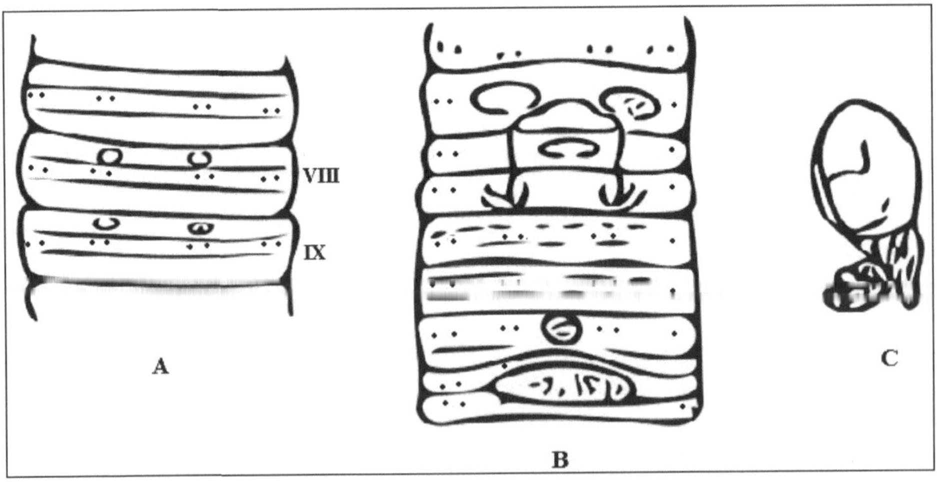

Figure 11.7: *Octochaetona compta* Gates, 1962: (A) Spermathecal pore region; (B) Genital region; (C) Spermatheca.

widened. Genital marking oval, unpaired and median on 13. Each spermatheca with dorso-ventrallly flattened and circular disc- like ectal diverticulum. Seminal vesicles 3 pairs in 9, 11 and 12.

Distribution: *Octochaetona* Andhra Pradesh.

Octochaetona albida (Figure 11.8) differs from the other *Octochaetona* species in having body length 53-96mm, body diameter 3-5mm, 154-185 body segments. Copulatory setae surrounded by oval to spindle shaped tumescences. Genital markings on 15/16/17; Genital marking glands absent. Each spermatheca fan shaped, shortly stalked, ectal diverticulum.

Distribution: Karnataka, Andhra Pradesh.

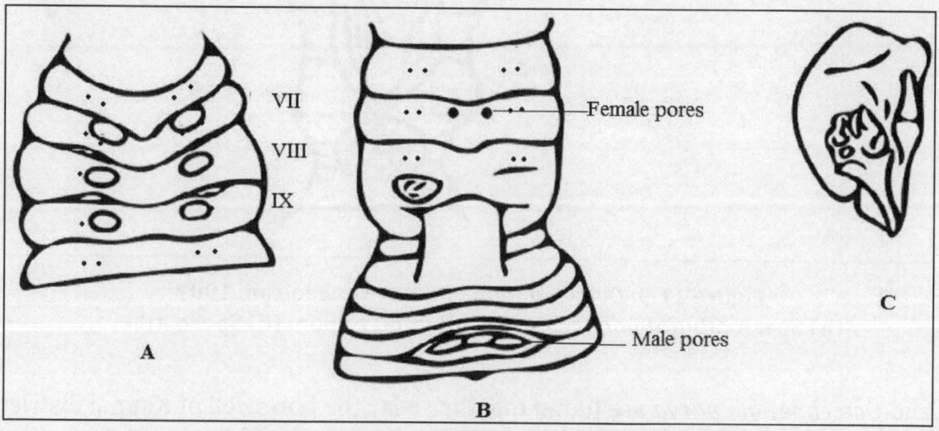

Figure 11.8: *Octochaetona albida* Gates, 1962:
(A) Spermathecal pore region; (B) Genital region; (C) Spermatheca.

Octochaetona phillotti (Figure 11.9) differs from the other *Octochaetona* species in having body length 93-128mm, body diameter 2-5.5mm, 157-199 body segments. Copulatory setae ornamented with transverse serrate ridges, tip claw shaped. Ciltellum annular; 13-16½. Pineal setae ornamented with circles of spines, tip bluntly pointed. Each spermathecae with self like ectal diverticulum, duct shorter than ampulla.

Distribution: Andhra Pradesh

All the *Octochaetona* species recorded in our study area during our study period are indigenous to India and endogeic, found at 10-20cm depth in our study area. Table 11.2 and Figure 11.10 depict the Habitat-wise distribution of *Octochaetona* species with their soil chemical parameters in the H-K region. Highest number of *Octochaetona* species were observed near bore well *i.e.*, 5 species; followed by near the gutter with 4 species, followed by in the garden with 4 species and least near the drainage system and the cultivated land with 4 species each. Association of *Octochaetona* species was observed in our study area. *Octochaetona paliensis*

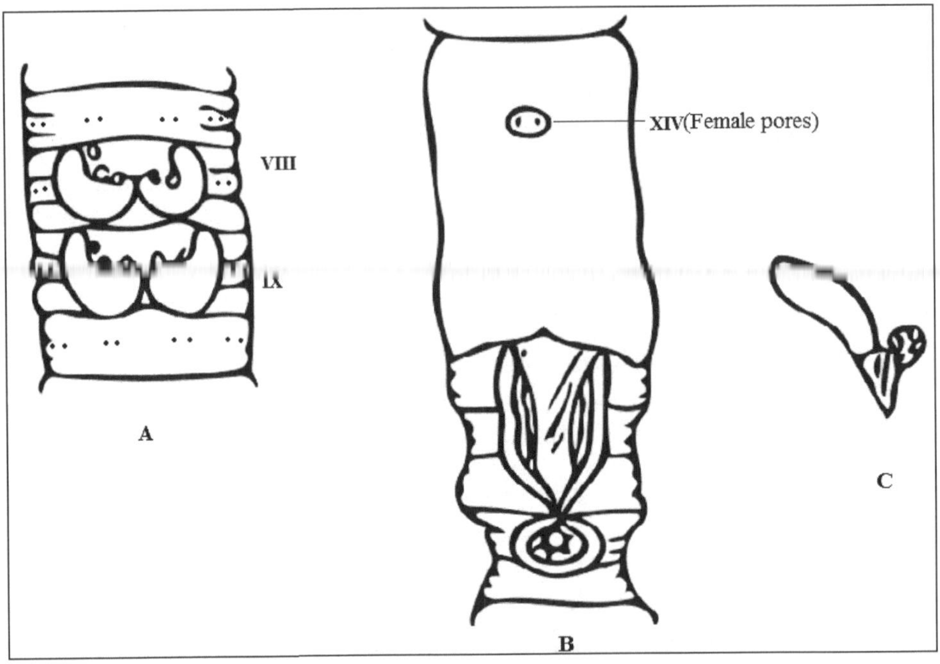

Figure 11.9: *Octochaetona phillotti* Michaelson, 1907: (A) Spermathecal pore region; (B) Genital region; (C) Spermatheca.

and *Octochaetona parva* are found together near the bore well of Koppal district of the H-K region.

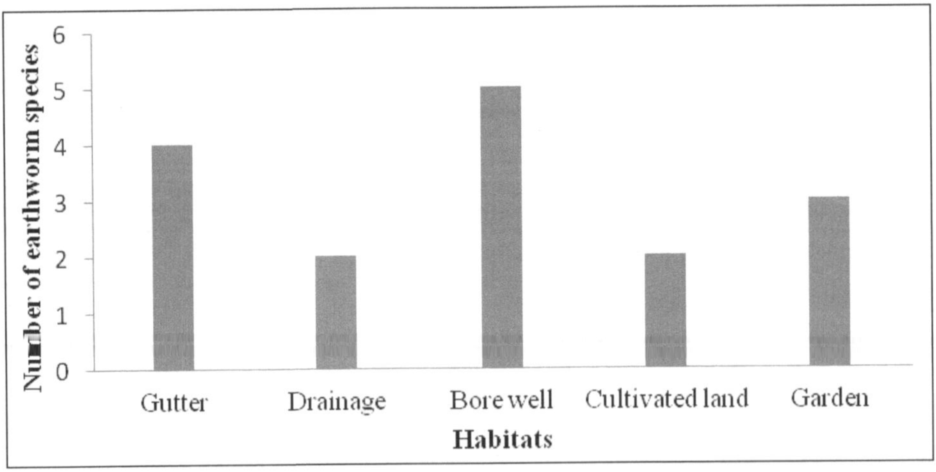

Figure 11.10: Distribution of *Octochaetona* Species in different Habitats of the H-K Region.

Table 11.1: The Details of the Morphological and Anatomical Characteristics of *Octochaetona* Species

Name of the Species	*Octochaetona paliensis Stephenson, 1920*	*Octochaetona beatrix Beddard, 1902*	*Octochaetona parva Gates, 1962*	*Octochaetona prashadi Gates, 1962*	*Octochaetona suriensis Michaelsen, 1910*	*Octochaetona compta Gates, 1962*	*Octochaetona albida Gates, 1962*	*Octochaetona phillotti Michaelson, 1907*
Body colour	Cream grey	Cream grey	Cream grey	Cream grey	Cream grey	Cream grey	Cream grey	Cream grey
Body length (mm)	35-92	40-134	70-100	51-61	60-140	70-150	53-96	93-128
Body Diameter (mm)	2-3.5	2-5	3-4	2.5-3.5	2.5-6	3-7	3-5	2-5.5
No. of Segments	119-182	133-197	80	150-155	111-180	131-176	154-185	157-199
Prostomium	Proepilobic	Epilobic	Epilobic	Combined protanylobic	Epilobic	Epilobic	Combined protanylobic	Epilobic
Setal arrangement	Lumbricine	Lumbricine	Lumbricine	Lumbricine	Lumbricine	Lumbricine	Lumbricine	Lumbricine
Spermathecal pores position and number	8 and 9; 2 pairs	8 and 9; 2 pairs	8 and 9; 2 pairs	7/8/9; 2 pairs	8 and 9; 2 pairs	8 and 9; 2 pairs	8 and 9; 2 pairs	8 and 9; 2 pairs
Copulatory setae description	Copulatory setae surrounded by tumescences	Ornamented with few scattered sigmoid excavations	Ornamented with longitudinal rows of fairly stout and long spines	Ornamented with longitudinal rows	Ornamented with longitudinal rows of spines, tip claw shaped	Ornamented with short transverse serrate lines	Copulatory setae surroundedby oval to spindle shaped tumescences	Ornamented with transverse serrate ridges, tip claw shaped

Name of the Species	***Octochaetona paliensis Stephenson, 192[illegible]***	***Octochaetona beatrix Beddard, 1902***	***Octochaetona parva Gates, 1962***	***Octochaetona prashadi Gates, 1962***	***Octochaetona suriensis Michaelsen, 1910***	***Octochaetona compta Gates, 1962***	***Octochaetona albida Gates, 1962***	***Octochaetona phillotti Michaelson, 1907***
Clitellum type and position	Annular; 13-17	Annular; 13-16	Annular; 13-17	Annular; 13-16	Annular; 13-16	Annular; 13-16	Annular; 13-17	Annular; 13-16$^1/_2$
Position and Number of Male pores	18; 1 pair	18; 1 pair	18; 1 pair	18; 1 pair	18; 1 pair	18; 1 pair	18; 1 pair	18; 1 pair
Seminal groove description	Present	aPresent	Present	Present	Present	Present	Present	Present
Position and number of prostatic pores	17 and 19; 2pairs	17 and 19; 2pairs	17 and 19; 2pairs	17 and 19; 2pairs	17 and 19; 2pairs	17 and 19; 2pairs	17 and 19; 2 pairs	17 and 19; 2pairs
Type of prostatic glands	Tubular	Tubular	Tubular	Tubular	Tubular	Tubular	Tubular	Tubular
Pineal setae description	Ornamented with circles of spines. Tip bluntly pointed	Ornamented with sparse triangular teeth tip	Ornamented with circles of fine spines, tip sharply pointed and bluntly rounded	Ornamented with closely set rings of spines, tip hooked	Ornamented with few longitudinal rows of triangular teeth, tip pointed or claw shaped	Ornamented with 15-30 circles of small teeth, tip sharp or rounded, narrowed or widened	Ornamented	Ornamented with circles of spines, tip bluntly pointed

Name of the Species	***Octochaetona paliensis* Stephenson, 1920**	***Octochaetona beatrix* Beddard, 1902**	***Octochaetona parva* Gates, 1962**	***Octochaetona prashadi* Gates, 1962**	***Octochaetona suriensis* Michaelsen, 1910**	***Octochaetona compta* Gates, 1962**	***Octochaetona albida* Gates, 1962**	***Octochaetona phillotti* Michaelson, 1907**
Position and number of Female pores	14; Paired	14; Paired	14; Paired	14; Paired	14; Paired	14; Paired	14; Paired	14; Paired
Position of first dorsal pores	12/13	12/13	12/13	12/13	12/13	12/13	12/13	12/13
Genital marking description	Absent	Absent	Absent	Absent	Absent	Oval, unpaired and median on 13	15/16/17; Genital marking glands absent	Absent
Position and number of Gizzards	4/5-7/8; Single	4/5-8/9; Single	4/5-7/8; Single	4/5-7/8; Single	4/5-8/9; Single	4/5-7/8; Single	6; Single	4/5-7/8; Single
Spermathecae description	Bidiver-ticulate	Each spermathecae with a spheroidal, shortly pyriform, oval or flattened and shelf-like, shortly stalked ental diverticulum	With disc like ectal diverticulum	Clavate ental diverticulum	Each spermathecae with shortly stalked, multiloculate ental diverticulum, duct longer than ampulla	Each spermatheca with dorso-ventrallly flattened and circular disc- like ectal diverticulum	Each spermatheca fan shaped, shortly stalked, ectal diverticulum	Each spermathecae with self like ectal diverticulum, duct shorter than ampulla

Name of the Species	***Octochaetona paliensis Stephenson, 192[illegible]***	***Octochaetona beatrix Beddard, 1902***	***Octochaetona parva Gates, 1962***	***Octochaetona prashadi Gates, 1962***	***Octochaetona suriensis Michaelsen, 1910***	***Octochaetona compta Gates, 1962***	***Octochaetona albica Gates, 1962***	***Octochaetona phillotti Michaelson, 1907***
Testis description	Holandric; 10 and 11	Metandric; 11	Holandric; 10 and 11	Holandric; 10 and 11	Holandric; 10 and 11	Holandric; 10 and 11	Holandric; 10 and 11	Holandric; 10 and 11
Seminal vesicle-position and number	9 and 12; 2 pairs	12; 1pair	11 and 12; 2 pairs	9 and 12; 2 pairs	9 and 12; 2 pairs	9,11 and 12; 3 pairs	9 and 12; 2 pairs	9 and 12; 2 pairs
Last pair of hearts	12	13	13	13	13	13	13	13
Beginning of Intestine	17	17	17	18	17	17	18	17
Typlosole	Present and ventrally bifid	Present and ventrally bifid	Present and ventrally bifid	Present and ventrally bifid	Present and ventrally bifid	Present and ventrally bifid	Present and ventrally bifid	Present and ventrally bifid

Soil pH

The mean values of soil pH of different habitats are represented in Table 11.2. The average values of pH is 7.9±0.89near the gutter,7.8±0.45near the bore well, 7.73±0.42in the garden, 7.55±0.21 in the cultivated lands and 7.675±0.60near the drainage system. The pH values recorded in our study are slightly alkaline *i.e.*, 7.55±0.21 to 7.9±0.89. The pH values are positively correlated with distribution of *Octochaetona* species. The relationship between the pH of the soil and distribution of *Octochaetona* species in different habitats is given in the Figure 11.11.

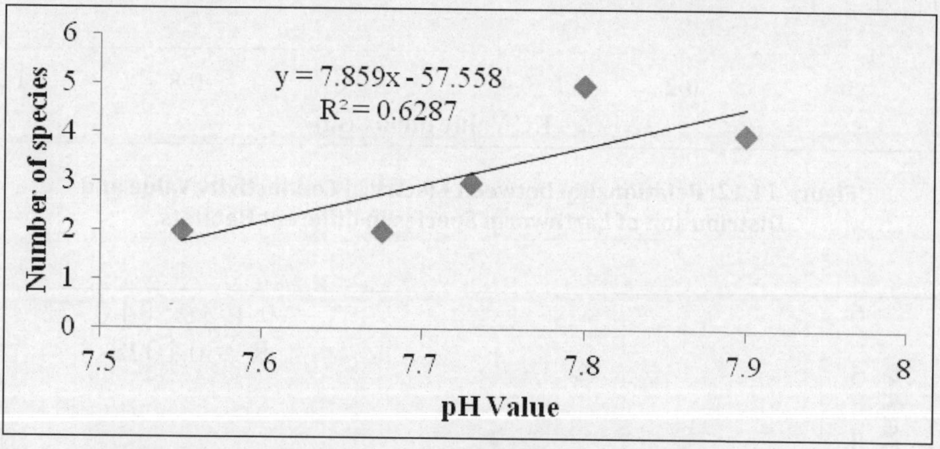

Figure 11.11: Relationship between pH Value and Distribution of Earthworm Species in different Habitats.

Electrical Conductivity

The mean values of soil EC of different habitats are represented in Table 11.2. The average values of EC is 0.74±0.26mohs/cm near the gutter, 0.39±0.15mohs/cm near the bore well, 1.38±0.60mohs/cm in the garden, 0.705±0.08mohs/cm in the cultivated lands and 0.26±0.03mohs/cm near the drainage system. The EC values are positively correlated with distribution of *Octochaetona* species. The relationship between the EC of the soil and distribution of *Octochaetona* species in different habitats is given in the Figure 11.12.

Organic Carbon

The mean values of soil OC of different habitats are represented in Table 11.2. The average values of OC is 1.95±0.65 per cent near the gutter, 1.43±1.04 per cent near the bore well, 1.38±0.60 per cent in the garden, 3.325±0.26 per cent in the cultivated lands and 1.12±0.40 per cent near the drainage system. The OC values are positively correlated with distribution of *Octochaetona* species. It was found to be highest in cultivated land, tilling and other factors did not affect them. The relationship between the OC of the soil and distribution of *Octochaetona* species in different habitats is given in the Figure 11.13.

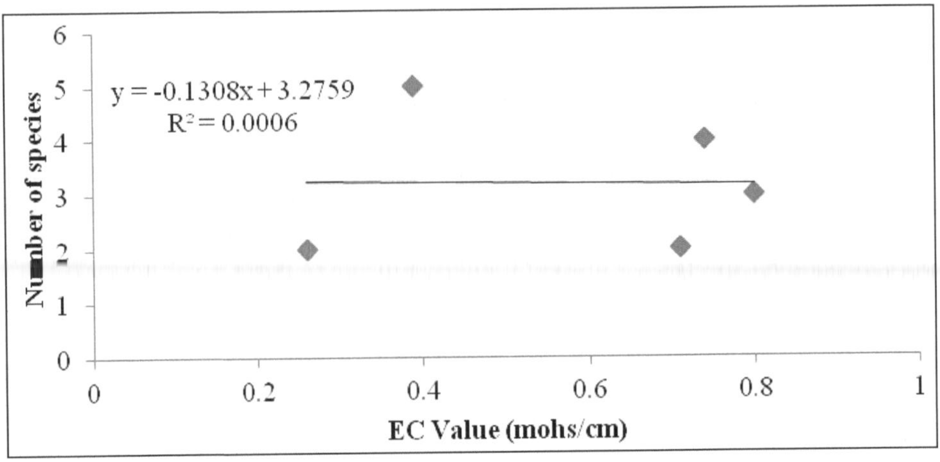

Figure 11.12: Relationship between Electrical Conductivity Value and Distribution of Earthworm Species in different Habitats.

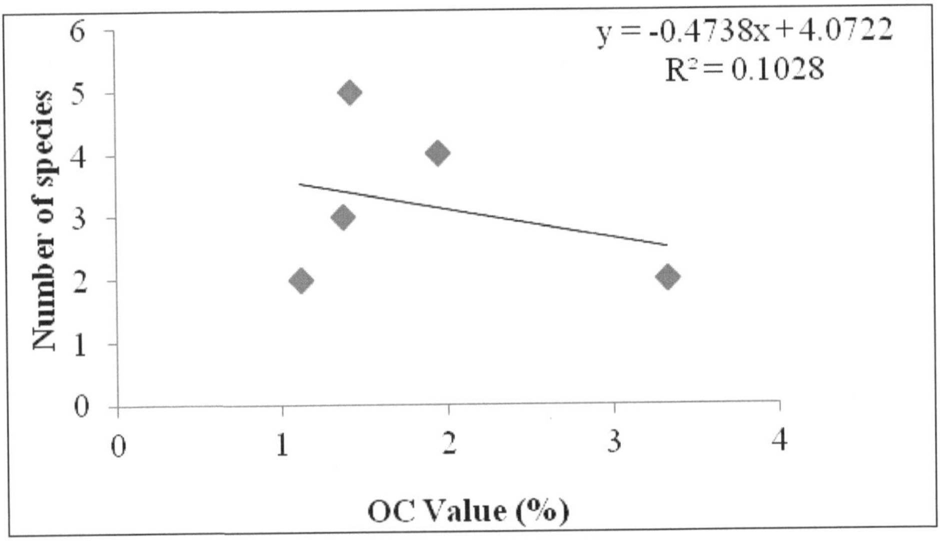

Figure 11.13: Relationship between Organic Carbon Value and Distribution of Earthworm Species in different Habitats.

Nitrogen

The mean values of soil nitrogen of different habitats are represented in Table 11.2. The average values of nitrogen is 65.25±21.71Kg/acre near the gutter, 47.85±34.77Kg/acre near the bore well, 46.23±20.03Kg/acre in the garden, 111.39±8.77Kg/acre in the cultivated lands and 37.52±13.27Kg/acre near the

Table 11.2: Habitat-wise Distribution of *Octochaetona* Species with their Soil Chemical Parameters in the H-K Region

Habitat	*Name of the Species*	*pH*	*EC (mohs/cm)*	*OC (per cent)*	*N (Kg/acre)*	*P (Kg/acre)*	*K (Kg/acre)*
Gutter	1. *Octochaetona paliensis* 2. *Octochaetona prashadi* 3. *Octochaetona suriensis* 4. *Octochaetonaphilloti*	7.9±0.89	0.74±0.26	1.95±0.65	65.25±21.71	25.59±14.66	234.45±74.70
Bore well	1. *Octochaetona paliensis* 2. *Octochaetona suriensis* 3. *Octochaetona beatrix* 4. *Octochaetona parva* 5. *Octochaetona prashadi*	7.8±0.45	0.39±0.15	1.43±1.04	47.85±34.77	23.16±11.74	251.08±68.85
Garden	1. *Octochaetona paliensis* 2. *Octochaetona compta* 3. *Octochaetona albida*	7.73±0.42	0.80±0.65	138±0.60	46.23±20.03	17.4±5.89	187±50.41
Cultivated lands	1. *Octochaetona paliensis* 2. *Octochaetona beatrix*	7.55±0.21	0.705±0.08	3.325±0.26	111.39±8.77	20.55±1.77	296±82.02
Drainage system	1. *Octochaetona beatrix* 2. *Octochaetona parva*	7.675±0.60	0.26±0.03	112±0.40	37.52±13.27	23.50±16.68	242.66±122.10

drainage system. The nitrogen values are negatively correlated with distribution of *Octochaetona* species. The relationship between the nitrogen of the soil and distribution of *Octochaetona* species in different habitats is given in the Figure 11.14.

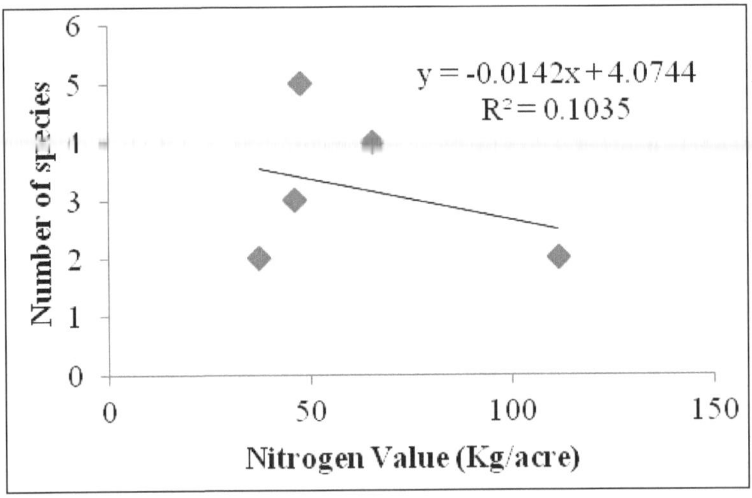

Figure 11.14: Relationship between Nitrogen Value and Distribution of Earthworm Species in different Habitats.

Phosphorous

The mean values of soil phosphorous of different habitats are represented in Table 11.2. The average values of phosphorous is 25.59±14.66Kg/acre near the gutter, 23.16±11.74Kg/acre near the bore well, 17.4±5.89Kg/acre in the garden, 20.55±1.77Kg/acre in the cultivated lands and 23.50±16.68Kg/acre near the drainage system. The phosphorous values are negatively correlated with distribution of *Octochaetona* species. The relationship between the phosphorous of the soil and distribution of *Octochaetona* species in different habitats is given in the Figure 11.15.

Potassium

The mean values of soil potassium of different habitats are represented in Table 11.2. The average values of potassium is 234.45±74.70Kg/acre near the gutter 251.00±60.05Kg/acre near the bore well, 187±50.41Kg/acre in the garden, 296±82.02Kg/acre in the cultivated lands and 242.66±122.10Kg/acre near the drainage system. The potassium values are positively correlated with distribution of *Octochaetona* species. The relationship between the potassium of the soil and distribution of *Octochaetona* species in different habitats is given in the Figure 11.16.

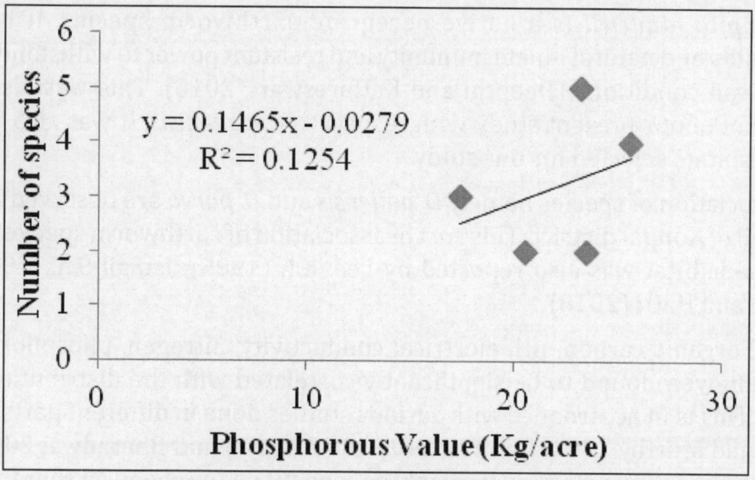

Figure 11.15: Relationship between Phosphorous Value and Distribution of Earthworm Species in different Habitats.

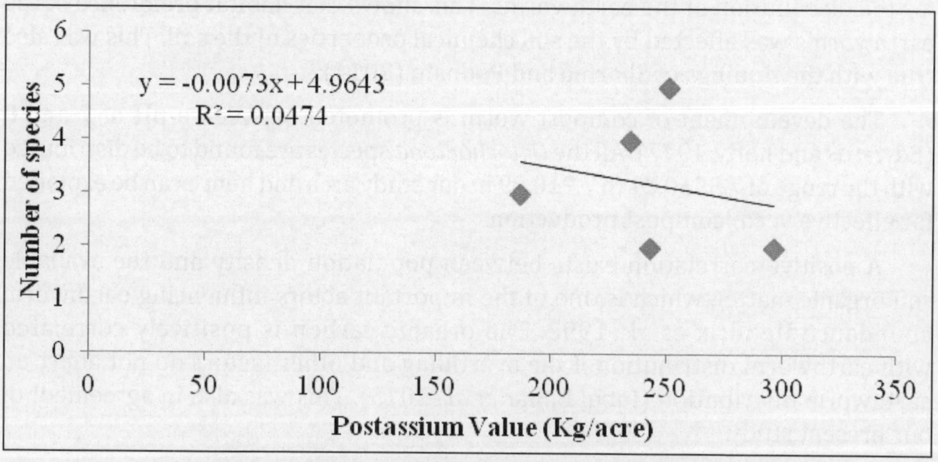

Figure 11.16: Relationship between Potassium Value and Distribution of Earthworm Species in different Habitats.

Discussion

It is the first random survey of earthworms in the H-K Region. *Octochaetona compta, O. paliensis, O. prashadi, O. suriensis, O. parva, O. beatrix O. albida* and *O. phillotti* are found in different habitats. This observation indicates that the distribution of different earthworm species is limited even though they are closely related. Such differences in the habitat preferences were also reported by other workers (Bhadauria and Ramakrishnan 1991; Chaudhuri and Bhattacharjee 1999; Kale and Karmegam 2010).

Lampito mauritii is a native peregrine earthworm species, it has high adaptability and natural innate immunity and resistant power to withstand against various soil conditions (Deepthi and Kathireswari, 2016). This was also in the agreement of our present study with respect to *O. palienisis*. It was also found in all the habitats selected for our study.

Association of species namely *O. paliensis* and *O. parva* are observed near the bore well of Koppal district. This sort of association of earthworm species sharing the same habitat was also reported by Lee K.E. (1985), Ismail S.A. (1986) and Mubeen and Hatti (2018).

Soil organic carbon, pH, electrical conductivity, nitrogen, phosphorous and potassium were found to be significantly correlated with the distribution of the worms. This is in accordance with various studies done in different parts in India (Ismail and Murthy 1991; Ganihar 1996 and Tripathi and Bharadwaj 2004). The distribution of earthworms was mainly dependent on the chemical characteristics of the soil. Soil pH in this region was alkaline (7.55±0.21 to 7.9±0.89). According to Edwards and Lofty (1977), earthworm species have narrow pH, very few restricted to highly acidic soil (pH<4). The pH values recorded were in the range for the distribution of the earthworms. This shows that habitat preference of the earthworms was affected by the soil chemical properties of the soil. This was also true with the findings of Sharma and Poonam (2014).

The development of compost worm is promoted between a pH of 7 and 8 (Edwards and Lofty, 1977). All the *Octochaetona* species are found to be distributed with the range of 7.55±0.21 to 7.9±0.89 in our study area and hence can be exploited for effective vermicompost production.

A positive correlation exists between population density and the available soil organic matter, which is one of the important actors influencing earthworm abundance Hendrix *et al.*, 1992. The organic carbon is positively correlated with earthworm distribution if the overtilling and other factors do not affect on earthworm distribution (Iqbal Bahar *et al.*, 2015). This was also in agreement of our present study.

There are more than a dozen of earthworm species that have been reported to be efficient in vermicomposting. Most of the species that are included under genus *Perionyx* show great potential to work on organic matter (Kale and Karmegam, 2010). Apart from the well known *Perionyx* Reddy and Okhura (2004) have claimed an increased level of per cent organic carbon in vermicompost transformed from rice straw using *Octochaetona phillotti*.

Conclusion

Eight species of the *Octochaetona* are identified from 23 sites of the H-K region. All the *Octochaetona* species recorded in our study area during our study period are indigenous to India and endogeic, found at 10-20cm depth in our study area. So far the occurrence of five *Octochaetona* species *i.e.*, *O. albida*, *O. beatrix*, *O. paliensis*,

O. prashadi and *O. parva* have been reported from Karnataka state. This is the first report of occurrence and distribution of *O. compta, O. philloti* and *O. suriensis* in Karnataka state. Vermicompost transformed from rice straw using *O. philloti* can increase level of per cent organic carbon Reddy and Okhura (2004).

Acknowledgment

1. The financial support provided by UGC in terms of Maulana Azad National Fellowship, New Delhi for work is gratefully acknowledged.
2. R. Paliwal ZSI Scientist (Retd,) for his exemplary guidance in training in earthworm taxonomy
3. Soil testing laboratory, Kotnoor, Kalaburagi

REFERENCES

Csuzdi CS.2012. Earthworm species, a searchable database. *pusc. Zool. Budapest,* 43(1): 97–99.

Dash, C., Madhab and K G Saxena, And. (2012). Earthworms in the Himalaya and Western Ghats region of India: a review. *An International journal of Life sciences* 7(1): 1-8.

Deepthi M. P. and Kathireswari P. 2016. Earthworm Diversity and Analysis of Soil Inhabited by Earthworms in the Vatakara area, Kozhikode, Kerala, India. *Int. J. Curr. Microbiol. App. Sci.* 5(3): 917-925

Gates, G.E. 1940. Indian earthworms. VIII-XI*Rec. Indian Mus.* 42: 115-143.

Gates, G.E. 1972. Burmese earthworms. An introduction to the systematics and biology of megadrile oligochaetes with special reference to Southeast Asia. Trans. Am. Phil. Soc. 62(7): 1-326.

Hatti, S.S. 2013. Taxonomical Studies on Earthworm Species of Gulbarga City, Karnataka, India. *Indian Journal of Applied Research*, 3: 34-39

Hendrix, P. J. B. R. Mueller and R.R. Bruce, "Abundance and distribution of earthworms in relation to landscape factors onthe Georgia Piedmont, USA." 1992, Soil Biol. Biochem., vol.24, pp. 1357-61.

Julka J. M. 1978. Studies on the earthworm fauna of Orissa (India). 2. Megascolecidae, Octochaetidae and Microchaetidae. *Mitt. Zool. Mus. Berlin*. 54: 185-197.

Julka J.M.1988. The fauna of India and adjacent countries: Megadrile Oligochaeta (Earthworms). Haplotaxida: Lumbricina: Megascolecoidea: Octochaetidae. *Zoological Survey of India*, Calcutta.xiv+400pp.

Julka J.M. 1995. Himalayan Ecosystem Series: Fauna of Western Himalaya.*Zoological Survey of India.* Part1: 17-22

Julka J.M and R Paliwal 1993. *Earthworms Resources and Vermiculture:* Zoological Survey of India.7-11.

Julka J.M and R. Paliwal. 2000. Oligochaeta, pp.21-25. In: Fauna of Renuka Wetland: Wetland ecosystem series 2. *Zoological Survey of India, Calcutta.*

Michaelsen, W. 1907. Neue Oligochaten von Vorder-Indian, Ceylon, Burma und den AndmanInseln. *Jb. hamb. wiss. Anst.* 24: 143-188.

Mohan V.C., Watts P. and Kaur A. 2013. Diversity and Distribution of Earthworms Species in Guru Nanak Dev University Campus, Amritsar, Punjab, India. *Research Journal of Agricultural and Forestry Sciences.* 1(2).pp 35-40

Muddaiah Siddaraju, Kanale S. Sreepada and Krishna M P, 2013. Recorded Distribution of Earthworms of the Family Octochaetidae In Dakshina Kannda District, South West Coast, Karnataka. *International Journal of Scientific Research Publications.* 3(6).

Om Prakash.2017. Biodiversity of Earthworms and their Distribution in Different Regions of Uttar Pradesh state of India. *IOSR Journal Of Pharmacy.* 7: 01-09.

Palanisamy Kathireswari, Jatinter M. Julka and John W. Reynolds.2005.Checklist of Oligochaeta of Tamil Nadu.*Megadrilogica.* 10(8).

Rahul Paliwal. 2008.Faunal Diversity of Jabalpur District, Madhya Pradesh. *Zoological Survey of India*. pp 29-53.

Radha D. Kale and R.V. Krishnamoorthy, 1978. Distribution and abundance of earthworms in Bangalore. *Proc. Indian Acad. Sci.* 87B: pp.23-25.

Ramzan Begum V.I. and Sultan Ahmed Ismail. 2003. Illustration of Earthworms occurring in and around Chennai, *India.Zoos' Print Journal* 19(3): 1394-1400.

Sharma R.K. and Bhardwaj Poonam. 2014. Earthworm Diversity in Trans-Gangetic Habitats of Haryana, India. Research *Journal of Agriculture and Forestry Sciences.* 2(2): 1-7.

Siddaraju M., K.S. Sreepada and J.W. Reynolds 2010. Checklist of Earthworms (Annelida: Oligochaeta) from Dakshina Kannada, Karnataka, Southwest India. *Megadrilogica.* 15 (5).

Srivastava, R., Kumar, M., Choudhary, A. K. and Sinha, M. P. 2003. Earthworm diversity in Jharkhand state. *J. Nat., Env. and Poll. Tech.* 2(3): 357-362.

Stephenson, J. 1923. Oligochaeta. The Fauna of British India, including Ceylon and Burma, Taylor Francis Ltd., London. pp. xxiv + 518.

Stephenson, J. 1925. On some Oligochaeta mainly from Assam, South India and the Andaman Islands. Rec. Indian Mus. 27: 43-73.

Tripathi. G and Bharadwaj. P. 2004. Earthworm diversity and Habitat preferences in arid regions of Rajasthan. *Zoos' print Journal* 19(7): 1515-1519.

Chapter 12

Spatial Heterogeneity of Phytoplankton Communities in Tungabhadra Impoundments, Hospet

C.M. Nagabhushan

Department of Studies in Zoology,
Vijayanagara Sri Krishnadevaraya University, Ballari, Karnataka
e-mail: nagabhushancm@vskub.ac.in

Aquatic ecosystems are rarely static, as their boundaries are subjected to change and fluctuate seasonally. Manipulation of reservoirs by setting up hydro-thermal plants, various types of industries, discharge of sludge and other commercio-domestic anthropocentric intervention disturbs the natural habitats. Tungabhadra reservoir located at coordinates 76° 21′ 10″ E and 15° 15′ 19″ N irrigates the agricultural land and generates hydro-electric power from 72 MW capacity plant. At the full level of 497.7 m above Mean Sea Level, the reservoir extends over 37,814 ha. The hydrological parameters were assessed in the decadal interval of time (2009-10 to 2018-19). The mean values of air temperature, water temperature, turbidity, TDS, TSS, Electric Conductivity, Total alkalinity, COD, BOD, Dissolved Oxygen, Sulphates, Nitrates, Phosphates, Calcium, Magnesium, Iron, Fluoride, Free carbon-dioxide, pH, Chloride and Total hardness that were compared with the same parameters approximately a decade ago. The results showed wide range of fluctuation in EC (38 to 342.50), increased BOD from 2.25 to 4.46 mg/l, high DO of 7.5mg/l, sulphates 25 to 65mg/l, less concentration of Nitrates from 2.00-3.05 mg/l, Calcium 17.5-43 mg/l, Magnesium 5.15-22.5 mg/l, relatively high levels of Iron (1.63-3.90 mg/l), high levels of Fluoride (0.4-1.2 mg/l), wide range of Carbon-dioxide (1.38-38 mg/l), alkaline pH from 7.5-8.58 and total hardness from 48-96.25±2. Four phytoplankton groups; Desmidaceae, Bacillariophyceae, Chlorophyceae and Cyanophyceae were enumerated at genus level. The immense water spread area of the reservoir and diluting nature of the lentic water ecosystem has retained all the hydrological parameters within the permissible limits of CPCB and BIS. The promising potentiality of Tungabhadra reservoir still attracts many fishermen for both captures and culture fishery.

Keywords: *Tungabhadra reservoir, Heterogeneity, Limnology, Phytoplankton.*

Introduction

Freshwater is a critical, finite, vulnerable renewable resource essential for life activities. The freshwater available in the country is useful to mankind in many respects *viz.*, potable, domestic, washing, bathing, agriculture and even for aquaculture. UN had declared the International year of freshwater to focus the issues of emerging water crisis globally way back in 2003. Aquatic ecosystems are rarely static, as their boundaries are subjected to change and fluctuate seasonally. Manipulation of reservoirs by setting up hydro-thermal plants, various types of industries, discharge of sludge and other commercio-domestic anthropocentric intervention disturbs the natural habitats.

The reservoir does not cause environmental hazards by converting them into eutrophic because of heavy loading of organic matter from the submerged land and runoff from the catchment area. Regular monitoring of abiotic and biotic factors has resulted in progress of productivity yet complete utilization of aquatic resources is not achieved since ages. The reservoirs are unique ecosystems; with varied variations in its quality and quantity caused by freshwater runoff from its tributaries and land drainage. Life in such environments is possible for those who adapt to the changing conditions.

There are over 1500 large reservoirs covering more than 1.45 million ha and more than one lakh medium and small sized reservoirs covering 1.1 million ha in India (Gopal, 1994). In Karnataka there are 12 large, 16 medium and 46 small sized reservoirs covering 223,887 ha of area and 4,605 tanks cover 213,404 ha of area. Total water area under man made impoundments cover 437,291 ha is undoubtedly one of the largest in the country holding tremendous potential for fisheries development. Study of plankton in freshwater is well established in Indian waters (Sugunan, 1991; Devaraj *et al.*, 1988; Sreenivasan, 1964, 1967; Trivedy, 1983; Pandey *et al.*, 2006; Hosetti 2009).

The present study explores the intricate relationship that exists between plankton and water quality prevailing in the lentic ecosystem. It also contributes to understanding the community level responses, sensitivity in the context of water quality gradient.

Materials and Methodology

The tributary of Krishna, Tungabhadra river covers eight districts from its source initiating from Chikamagalore, Shimoga, Davanagere, Haveri, Bellary,

Vijayanagara, Koppal and Raichur; twenty eight taluks namely Magalore, Tarikere, Koppa, Sringeri, N.R.Pura, Shimoga, Bhadravathi, Channagiri, Harihar, Honnali, Ranebennur, Hirekerur, Haveri, Hospet, Siruguppa, Bellary, Koppal, Gangavathi, Sindhanoor, Manvi and Raichur in Karnataka and four districts in Andhra Pradesh; Mehaboob Nagar, Kurnool, Ananthpur and Cuddappah. This basin is mostly rainfed, dominated by red soils. Agriculture is the major occupation and major crops grown are paddy, jowar, sugarcane, cotton and millet.

Topographically three different sampling stations (S1, S2 and S3) were chosen in the shallow zone ensuring that the selected zone always retained water throughout all the seasons even during the draught located in the coordinates at 76° E and 15° N.

The water samples were collected from the peripheral area at 1 to 1.5 meter depth in the one liter wide-mouthed-plastic-cans and transported to the laboratory for analysis. Air and water temperature were recorded at the sampling site itself at 9.00 a.m. However due to lack of boat facility collection in the middle of the water body was not made. The pH of water body was also recorded at the sampling sites using pH meter. Further analysis of water parameters such as free CO_2, Dissolved Oxygen (DO), Total alkalinity (TA), Total Dissolved Solids (TDS), Total Suspended Solids (TSS), Phosphates, Sulphates, Nitrates, Total hardness, Electrical conductivity, Fluoride, Chloride, Turbidity, Iron, Magnesium, Calcium including BOD and COD were carried out immediately as per the standard methods (APHA, 1998, 2000). The plankton samples were collected using the plankton net with the mesh size, 50μm. 100 liters of water was filtered through the plankton net fitted with 100 ml plankton collecting bottle attached to the bottom of the net. The plankton sample was then fixed in 4 per cent formaldehyde solution or Lugol's Iodine and left for sedimentation. After sedimentation random samples were siphoned out to put in the plankton counting chamber for enumeration. The enumerated planktonic observation was presented in percentage compositions. The results obtained were represented as units/l for phytoplankton and organism/l for zooplankton. The Phytoplankton were identified upto genera level whereas zooplankton were identified upto species level. The quantitative analysis was performed using the formula (Santhanam *et al.*, 1989);

$$N = \left[\frac{n\ X\ v}{V}\right] X\ 1000$$

where,

N = Total number of plankton per liter of water filtered

n = Mean number of plankton in 1 ml of plankton sample

v = Volume of plankton sample concentration in ml

V = Volume of total water filtered in litres.

Results and Discussion

The surface water temperature ranged from 22°C-30°C in S1, 25°C-30°C at S2, 25°C-30.5°C at S3, respectively. It was between 25°C to 30°C during 2010-11 (Nagabhushan, 2011). Similar recordings were observed by Dynaneshwari Patil and Meena Dongre (2006) in lentic water bodies of Kolhapur, Maharashtra. The presence of dissolved substances absorbed light and causes a negative interference thereby affecting biota. High mean value of turbidity recorded was 120.50 NTU during summer 2009 and lowest of 11.5 during winter in 09. Usha *et al.* (2006) also reported high turbidity in April month in the Perumal Lake, Cuddalore, Tamil Nadu. High content of TDS reduces the light penetration and affect water quality indirectly and affects the growth of live plankton and thereby fishes. In the present study maximum solids were recorded in the rainy months. Thus high values of TDS in monsoon can be correlated with the leaching of soil and silt carried out into the reservoir from the catchment area and minimum values in the winter are attributed to intense sun radiation in the area resulted in evaporation of water. The results showed wide range of fluctuation in EC (38 to 342.50), increased BOD from 2.25 to 4.46 mg/l, high DO of 7.5 mg/l, Sulphates 25 to 65 mg/l, less concentration of Nitrates from 2.00 - 3.05 mg/l, Calcium 17.5 - 43 mg/l, Magnesium 5.15 - 22.5 mg/l, relatively high levels of Iron (1.63 - 3.90 mg/l), high levels of Fluoride (0.4 - 1.2 mg/l), wide range of Carbon-dioxide (1.38 - 38 mg/l), alkaline pH from 7.5 - 8.58 and total hardness from 48 - 96.25 ±2. The seasonal mean values of all the hydrological parameters are depicted in the Table 12.2. Kiran *et al.* (2007) have reported similar plankton groups. SHE analysis examines the relationship between Species richness (S), Shannon Weiner diversity index (H') and Shannon Weiner evenness index or Pielou J (E) in the samples. It is an approach to look at the contribution of species number and equitability to changes in diversity. SHE analysis follows the way the parameters changes with increasing sampling efforts. Hayek and Buzas (1997) opined that SHE analysis was useful for identifying ecotones of a region where different ecological communities intersect. Species richness seemed to be high and least evenness as well as less similarity index of phytoplankton groups in all the stations.

In the present investigation 9 genera and 13 species of Desmids were encountered during 2009-10 but the representation during 2018-19 was 9 genera and 11 species. The species composition is represented by *Closterium lineatum, Closterium moniliferum, Cosmarium ochthodes, Euastrum binale, Gonatozygon aculeatum, Microsterias anomola, Microsterias mahabalipuriensis, Microsterias thomasiana, Penium sp., Pleurotaenium trabecula, Spondylosium planum* and *Staurastrum altermans* dominated by *Microsterias anomola, M. mahabalipuriensis, M. Thomasiana.* Zafar (1967) opined that alkaline pH favoured the abundance of desmids. It was in agreement with the present findings.

Chlorophyceae was represented by 15 genera and 16 species during 2009-10 and during 2018-19 there were 14 genera and 14 species. The diversity was represented by *Ankistrodesmus falcatus, Bumilleria sp., Chaetophora incrassata,*

Cylindrocystis brebisonii, Hormidium flaccidum, Mougeotia calcarea, Netrium digitus, Oedogonium sp., *Pediastrum duplex, Scenedesmus bijunga, Scenedesmus obtusus, Sphaerozosma granulatum, Spirogyra inflata, Stigeoclonium helveticum, Tetraspora* sp. and *Zygnema cylindricum.*

Table 12.1: Standards for the Drinking Water Quality

Sl.No.	*Parameters*	*Unit*	*Permissible limit*	*Excessive limit*	*Status*
1	Colour	Hazen	5	50	–
2	pH	–	7.0 - 8.5	6.5 - 9.2	#
3	Turbidity	NTU	5	25	#
4	Total Dissolved Solids	mg/L	500	2000	#
5	Electrical Conductivity	µS/cm	1000	2250	#
6	Chloride	mg/L	200	1000	#
7	Total hardness	mg/L	200	600	#
8	Dissolved Oxygen	mg/L	4.00	6.00	#
9	Biochemical Oxygen Demand	mg/L	2.00	3.00	#
10	Fluoride	mg/L	1.00	1.50	#
11	Calcium	mg/L	75	200	#
12	Magnesium	mg/L	30	150	#
13	Sulphate	mg/L	200	400	#
14	Nitrate	mg/L	45	45	#
15	Total Alkalinity	mg/L	200	400	#
16	Iron	mg/L	0.30	1.00	#
17	Manganese	mg/L	0.05	0.5	–
18	Copper	mg/L	0.05	1.5	–
19	Zinc	mg/L	5	15	–
20	Phenolic Compounds	mg/L	0.001	0.002	–
21	Arsenic	mg/L	0.05	0.05	–
22	Cadmium	mg/L	0.01	0.01	–
23	Chromium	mg/L	0.05	0.05	–
24	Lead	mg/L	0.10	0.10	–
25	Cyanides	mg/L	0.05	0.05	–
26	Selenium	mg/L	0.01	0.01	–
27	Mercury	mg/L	0.001	0.001	–

#: Parameter tested; –: Parameter not tested during study period.

Courtesy: Bureau of Indian Standard– 14543:2004; IS – 13428:2005.

Table 12.2: Seasonal Mean Values of Hydrological Parameters in Three Stations during 2009–10 and 2018–19

Hydrological Parameters	*2009-10*			*2018-19*		
	Rainy	*Winter*	*Summer*	*Rainy*	*Winter*	*Summer*
Air temperature	28.75	29.38	31.13	29.25	29.23	30.60
Water temperature	28.35	28.88	29.63	28.55	28.50	29.75
Turbidity	15.25	13.75	15.25	15.00	15.75	12.50
TDS	215.00	165.00	247.50	210.00	220.00	237.50
TSS	87.50	50.00	62.50	85.00	60.00	62.50
Electrical Conductivity	342.50	230.00	275.00	305.00	270.00	270.00
Total Alkalinity	65.00	93.75	85.00	82.50	90.00	97.50
COD	87.50	322.50	102.50	147.50	272.50	82.50
BOD	2.75	3.00	3.25	2.50	3.25	2.25
Dissolved Oxygen	4.93	5.93	5.68	5.75	5.50	5.13
Sulphate	42.50	37.50	42.50	65.00	60.00	45.00
Nitrate	2.35	2.48	2.23	2.50	2.55	2.05
Phosphate	2.50	2.00	1.25	2.75	3.50	1.75
Calcium	30.75	27.25	31.75	28.75	30.00	33.75
Magnesium	10.08	7.13	11.28	11.13	8.50	11.05
Fe (Iron)	3.25	2.13	1.88	2.63	1.63	2.50
Fluoride	0.60	0.58	0.45	0.53	0.58	0.68
Free Carbon-dioxide	1.63	1.43	1.65	1.60	1.43	1.58
pH	8.08	8.30	8.38	8.25	8.40	8.48
Chloride	38.50	30.25	33.25	35.00	31.25	23.75
Total Hardness	86.25	75.00	91.25	89.00	79.50	98.00

Values are in mg/L, except pH and for air temperature and water temperature in °C, turbidity=NTU, EC= µS/cm.

Cyanophyceae members consisted of a total of 11 genera and 18 species during 2009-10 and in 2018-19. They included, *Acanthocystis nichollsi, Anaebena aequalis, Aphanocapsa koordersi, Aphanocapsa biformis, Coelesphaerium kuetzingianum, Lyngbya martensiana, Merismopedia minima, M. punctata, Microcystis aeruginosa, M. marginata, M. viridis, Oscillatoria curviceps, O. princeps, O. tenuis, Phormidium acutum, Spirulina laxissima, S. platensis* and *Stigonema ocellatum*

The Bacillariophyceae presented 18 genera and 21 species. *viz., Achranthes* sp., *Amphora montanta, Asterionella* sp., *Coscinodiscus centralis, Coscinodiscus lineatus, Cyclotella meneglicana, Cymbella* sp., *Diatoma vulgare, Diploneis* sp., *Fragillaria construens, Gomphonema constrictum, Gyrosigma kutzingii, Melosira varians, Navicula oblonga, Nitzschia palae, Pinnularia viridis, Pleurosigma angulatum,*

Surirella angusta, Synedra acicularis and *Synedra ulna.* During 2018-19 the Bacillariophyceae dominated the planktonic population.

Table 12.3: Percentage Composition of Phytoplankton Community over the Study Period

Phytoplankton Family	*2009-10*	*2018-19*
Desmidaceae	20.22	21.28
Chlorophyceae	21.60	22.51
Cyanophyceae	24.73	21.64
Bacillariophyceae	33.45	34.57
Total per cent	100	100

Conclusion

TBR is a joint venture operation of Hyderabad-Karnataka, yearly scientific documentation of the entire TBR is thin since impoundment of the dam. The low density of plankton during rainy season was due to turbid water, which reduced penetration of sun light and adversely affects the production of plankton. A similar observation was reported by Sugunan *et al.* (2000). In the present study alkaline pH, low levels of Calcium, Phosphate and Turbidity promoted the growth of Desmids. Continuous monitoring of water is required to know the status of TungaBhadra reservoir. It has suggested a possibility of prediction for desmid variability patterns by some water quality parameters. The scientific appraisal of plankton community provides information necessary for understanding of growth and abundance of fin-fish population. In the present study abundance and diversity of four major phytoplanktonic groups were represented by Desmidaceae, Chlorophyceae, Cyanophyceae and Bacillariophyceae are shown in Table 12.4.

Table 12.4: List of Phytoplankton Species Recorded from Tunga Bhadra Reservoir

	Family: DESMIDACEAE	12	*Spondylosium planum*
1	*Closterium lineatum*	13	*Staurastrum altermans*
2	*Closterium moniliferum*		**Family: BACILLARIOPHYCEAE**
3	*Cosmarium ochthodes*	14	*Achranthes* sp.
4	*Euastrum binale*	15	*Amphora montana*
5	*Euastrum verrucosum*	16	*Asterionella* sp.
6	*Gonatozygon aculeatum*	17	*Coscinodiscus centralis*
7	*Micrasterias anomola*	18	*Coscinodiscus lineatus*
8	*Micrasterias mahabalipurensis*	19	*Cyclotella meneglicana*
9	*Micrasterias thomasiana*	20	*Cymbella* sp.
10	*Penium* sp.	21	*Diatoma vulgare*
11	*Pleurotaenium trabecula*	22	*Diploneis* sp.

23	*Fragillaria construens*	46	*Oscillatoria curviceps*
24	*Gomphonema constrictum*	47	*Oscillatoria principes*
25	*Gyrosigma kutzingii*	48	*Oscillatoria tenuis*
26	*Melosira varians*	49	*Phormidium acutum*
27	*Navicula oblonga*	50	*Spirulina laxissima*
28	*Nitzschia palae*	51	*Spirulina platensis*
29	*Pinnularia viridis*	52	*Stigonema ocellatum*
30	*Pleurosigma angulatum*		**Family: CHLOROPHYCEAE**
31	*Pleurosigma strigosum*	53	*Ankistrodesmus falcatus*
32	*Surirella angusta*	54	*Bumilleria* sp.
33	*Synedra acicularis*	55	*Chaetophora incrassata*
34	*Synedra ulna*	56	*Cylindrocystis brebisonii*
	Family: CYANOPHYCEAE	57	*Hormidium flaccidum*
35	*Acanthocystis nichollsi*	58	*Mougeotia calcarea*
36	*Anaebena aequalis*	59	*Netrium digitus*
37	*Aphanocapsa biformis*	60	*Oedogonium* sp.
38	*Aphanocapsa koordersi*	61	*Pediastrum duplex*
39	*Coelosphaerium kuetzingianum*	62	*Scenedesmus bijunga*
40	*Lyngbya martensiana*	63	*Scenedesmus obtusus*
41	*Merismopedia minima*	64	*Sphaerozosma granulatum*
42	*Merismopedia punctata*	65	*Spirogyra inflata*
43	*Microcystis aeruginosa*	66	*Stigeoclonium helveticum*
44	*Microcystis marginata*	67	*Tetraspora* sp.
45	*Microcystis viridis*	68	Zygnema cylindricum

In all, 68 different species of phytoplankton were recorded during 2009-10 and 60 during 2018-19. SHE analysis depicted high species richness and less similarity index and least evenness (Figures 12.1 and 12.2). The hierarchical cluster analysis indicated 97.89 per cent of similarity across the decade. Fish are the main keystone species which determine the distribution and abundance of other organism in the aquatic ecosystem they represent and are good indicators of water quality and aquatic ecosystem (Phun Kunwar Singh Rana, 2016). Any fluctuation in the microbiota, plankton biomass and diversity will reflect upon the zooplankton community thereby affecting the fish production as certain plankton are the indicators of water quality.

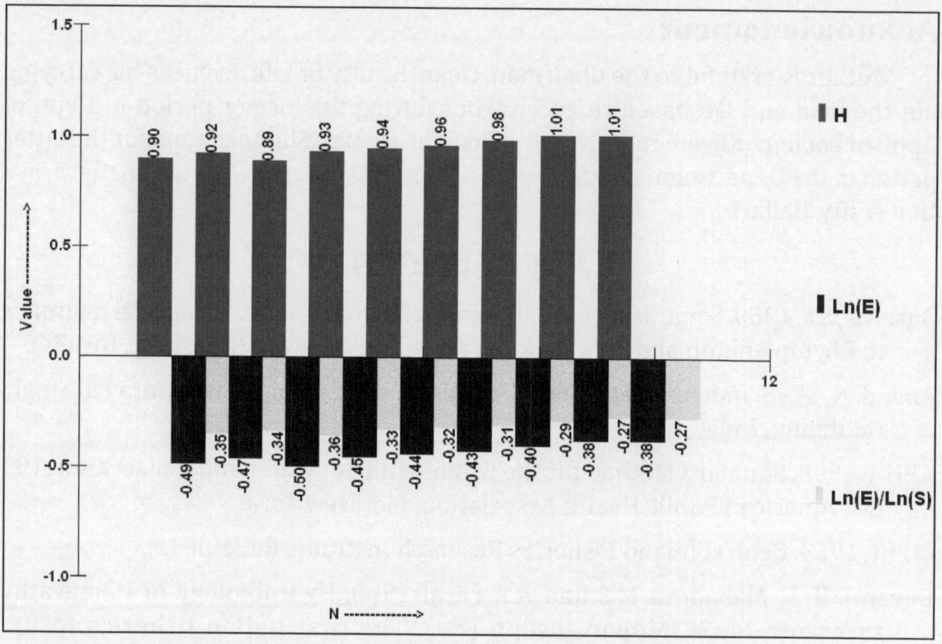

Figure 12.1: SHE Analysis Showing Species Richness, Shannon Weiner Diversity Index and Evenness Index of Phytoplankton during 2009-10.

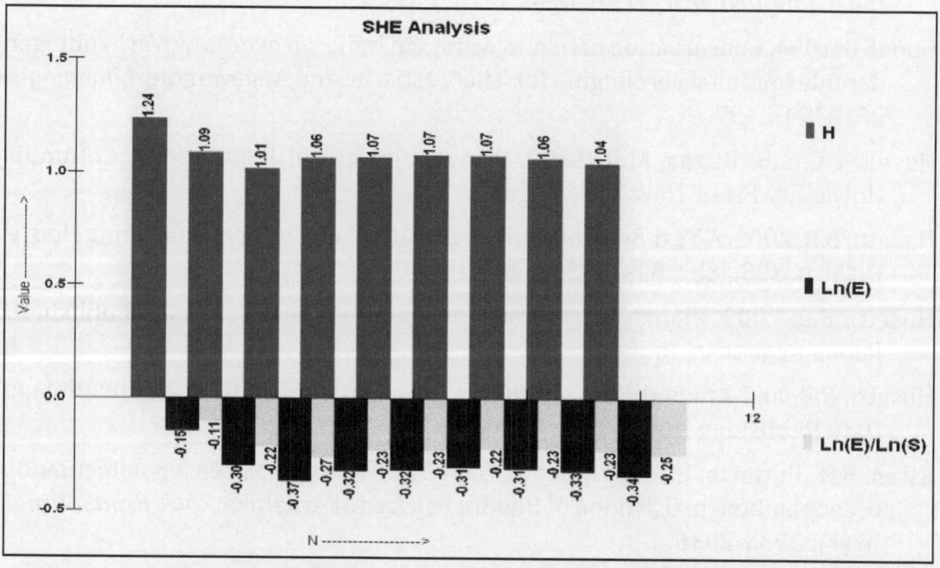

Figure 12.2: SHE Analysis Showing Species Richness, Shannon Weiner Diversity Index and Evenness Index of Phytoplankton during 2018-19.

Acknowledgment

Author is grateful to the Chairman, Dean, faculty of Life Sciences for carrying out the field and lab-based research work during the former period in Dept. of Applied Zoology, Kuvempu University, Shankaraghatta, Shimogga and for the latter period in the Department of Studies in Zoology, Vijayanagara Sri Krishnadevaraya University, Ballari.

REFERENCES

Agarwal, S.S. 1980. Some aspect of Limnology of Ramaua dam with special reference to Phytoplankton and Zooplankton. Ph.D. Thesis. Jiwaji University, Gwalior.

Anand, N. 1998. Indian Freshwater Microalgae. Bishen Singh Mahendra Pal Singh, Dehradun, India.

APHA. 1998. Standard Method for the Examination of Water and Wastewater. 19th Ed., American Public Health Association, Inc., New York.

CIFRI. 1963. Central Inland Fisheries Research Institute. Bulletin 13.

Devaraj. K. V, Mahadeva M.S and A.A Fazal 1988. Hydrobiology of Hemavathy reservoir. In: M. Mohan Joseph (Ed). The first Indian fisheries forum proceedings, Asian Fisheries Society, Indian Branch, Mangalore P. 323-327.

Dynaneshwari Patil and Meena Dongare. 2006. Seasonal variation in dissolved oxygen and biochemical oxygen demand of some lentic fishpond at Bhagalpur, India. *Env. and Ecol.*, 8(3): 929-932.

Gopal, B. 1994. Conservation of inland waters in India: an overview. Verhandlugen der international vereiniging fur Theorestische and Angewandie. *Limnologie.* 25: 2494-2497.

Hayek, L.C. and Buzas, M.A. 1997. Surveying Natural Populations. Columbia University Press, New York, p. 563.

Hosetti, B.B. 2001. A Text Book of Applied Aquatic Biology. Daya Publishing House, Delhi, p. 393, ISBN 81-7035-239-8.

Hosetti, B.B. 2009. Plankton Dynamics of Indian Waters. Pratiksha Publishers, Jaipur, Rajasthan, India.

Hosetti, B.B. and Aravind Kumar. 2001. A Text Book of Applied Aquatic Biology. Daya Publishing House, New Delhi.

Kiran, B.R., Puttaiah, E.T and Kamath, D. 2007. Diversity and seasonal fluctuation of zooplankton in fish pond of Bhadra fish farm, Karnataka. *Zoos' Print Journal,* 22(1): 2935-2936.

Nagabhushan C.M and Hosetti. B.B, 2012. The effect of climatic change on the diversity and composition of phytoplankton community in Tungabhadra reservoir.*International Journal of Scientific research and publications.* Volume 2, Issue 2, February 2012 Edition [ISSN 2250-3153]

Pandey, P.K., Jha, B.C., Sharad D. Patil and Som Dutt. 2006. Comparison of plankton populations of two aquatic habitats using biodiversity index. *J. Aquatic Biol.,* 21(1): 14-18.

Sreenivasan, A. 1964. The limnology, primary production and fish production in tropical pond. *Limnol. Oceanogr.,* 9(3): 391-396.

Sreenivasan, A. 1967. Application of limnological and primary productivity studies in fish culture. *FAO Fish Repo.,* 44(3): 101-113.

Sugunan, V.V. 2000. Ecology and fishery management of reservoirs in India. *Hydrobiologia,* 430: 121-147.

Sugunan, V.V. 1995. Reservoir Fisheries of India. FAO. Fisheries Tech. Report. 345. Daya Publishing House, New Delhi, India.

Sugunan, V.V. 2000. Reservoir Fishery Management in India (In Fifth Indian Fisheries Forum). Central Institute of Freshwater Aquaculture, Bhubaneshwar, India, pp. 104-111.

Tonapi, G.T. 1980. Freshwater Animals of India in Ecological Approach. Oxford and IBH Publ. Co., New Delhi, pp. 314-315.

Trivedi, R.K. and Goel, P.K. 1984. Chemical and Biological methods for water pollution studies. *Environmental Publication,* Karad, pp. 244.

Usha, R., Ramalingam and Raja, B. 2006. Freshwater lakes- A potential source for aquaculture activities – A model study on Perumal Lake, Cuddalore (Tamil Nadu), India. *J. Environ. Biol.* 27(4): 713-722.

WHO. 1984. International Guideline for the Drinking Water Quality. Vol. 1 and 2, Health criteria and other supporting information, Geneva.

Zafar, A.R. 1967. On the ecology of certain fish ponds of Hyderabad, India III. The periodicity. *Hydrobiol.,* 30(1): 96-112.

Chapter 13

Preliminary Study on Beetle Diversity (Insecta : Coleoptera) from Ballari Taluk, Karnataka

Sneha Suma Hegde

Department of Studies in Zoology,
Vijayanagara Sri Krishnadevaraya University, Ballari – 583 104, Karnataka
e-mail: snehasuma@vskub.ac.in

A preliminary survey was carried out from August 2019 to January 2020 in agricultural fields, grassland areas, crop lands, water bodies, herbs, shrubs, and open fields to know the diversity of beetles. A total of 16 species belonging to 08 families viz. Carabidae, Scarabaeidae, Tenebronidae, Buprestidae, Curculionidae, Chrysomelidae, Elateridae and Hydrophilidaewere recorded. Among these family Scarabaeidae shares maximum species (05),Carabidae shares (03) species, Tenebronidae and Curculionidae share each (02) species. The families Buprestidae, Chrysomelidae, Elateridae and Hydrophilidae had single species each. This survey is the first attempt to study the beetle diversity in Ballari region. This study will provide the basic data for researchers who have special interests in the conservation of beetles in this region.

Keywords: *Beetles, Coleoptera, Diversity, Ballari.*

Introduction

Insects numerically make up a major part of biodiversity (Erwin 1982, May 1988, Wilson 1992). Beetles are the foremost common variety of insects; however beetles are confused with other forms of insects, particularly some true bugs. Most insects have wings, in two pairs. Beetles differ from all other winged insects by having the primary pair of wings hardened and thickened into a sheath-like or shell-like protection for the rear part of the beetle's body. Coleoptera, regarded as the largest order of insects, has an estimated 350,000 species of beetles that are classified into 160 families (Elzinga 1992). Coleoptera is an order of insects commonly called beetles. The word "coleoptera" is from the Greek keleos, meaning "sheath," and pteron, meaning "wing," thus "sheathed wing" (Moitreyee Banerjee, 2014). The order Coleoptera includes more species than any other order, constituting almost 25 per cent of all known life-forms (J. A. Powell, 2009., M. L. Rosenzweig, 1995., T. Hunt *et al.,* 2007).Insects especially the beetles (Order: Coleoptera) was studied throughout the world on large basis. Over one and a half million living and about 12,000 species of fossil insects have been identified and described all over the world (Uniyal, 2001). About 1, 5088 species of beetles are known from Indian region (Kazmi and Ramamurthy, 2004).Beetles can both damage and help the surrounding environment. Some beetle species destroy crops or property; while some species help get rid of garbage and help pollinate flowers. Beetles are a group of mostly predatory insects, abundant in the field; their coloration, shape, and activity attract people in general. They are nocturnal hunters and prey on a wide range of small animals such as other insects and spiders; some species are diurnal and feed on plant tissues (Shirbhate Milind *et al.,* 2020). Several workers have reported beetle species in different areas (Choate, 2001). Beetles play necessary ecological role within the forest ecosystems in initiating and promoting the decay process of trees often acting as vectors of pathogenic or decomposer fungi (Crowson 1981). Several beetles are considered as major pests of agricultural plants and stored products. They attack all parts of living plants as well as processed fibers, grains, and wood products. Scavengers and wood boring beetles are helpful as decomposers and recyclers of organic nutrients. Predatory species, such as lady beetles, are important biological control agents of aphids and scale insects. Some beetles also act as food for fishes in fish culture practices. They are the pollinating agents in some crop plants. Beetles degrade the cattle dung and dung of forest dwelling animals like elephant, horse, *etc.* (D. S. Dabhade *et al.,* 2012).India is rich in terms of biological diversity due to its unique biogeographic location, varied climatic conditions and enormous eco-diversity (Satheesha *et al.,* 2018). India contributes to a diversified fauna of beetle and holds about 5 per cent of all known species of the world (Alfred JRB *et al.,* 1998). Beetles are considered as biological markers to assess the effects of anthropogenic activities on the terrestrial ecosystem, because they are in close contact with toxic elements present in soil and in leaf litter. These organisms are extremely sensitive and react quickly to environmental changes. These criteria make beetles as an excellent indicator of terrestrial ecosystem (Ghannem S, *et al.,* 2018). The present work focuses on the diversity of beetles in Ballari taluk.

Materials and Methods

Area of study

In the present investigation to study the beetle diversity in Ballari taluk the geographical regions selected were the agricultural fields, crop lands, water bodies, herbs, shrubs, forest trees and open fields. Bellary taluk is situated at coordinate 15.1394°N Latitude and 76.9214°E Longitude. The temperature and humidity recorded during the study were 31°C and 23 per cent. The average rainfall in this region is about 639 mm (Figure 13.1).

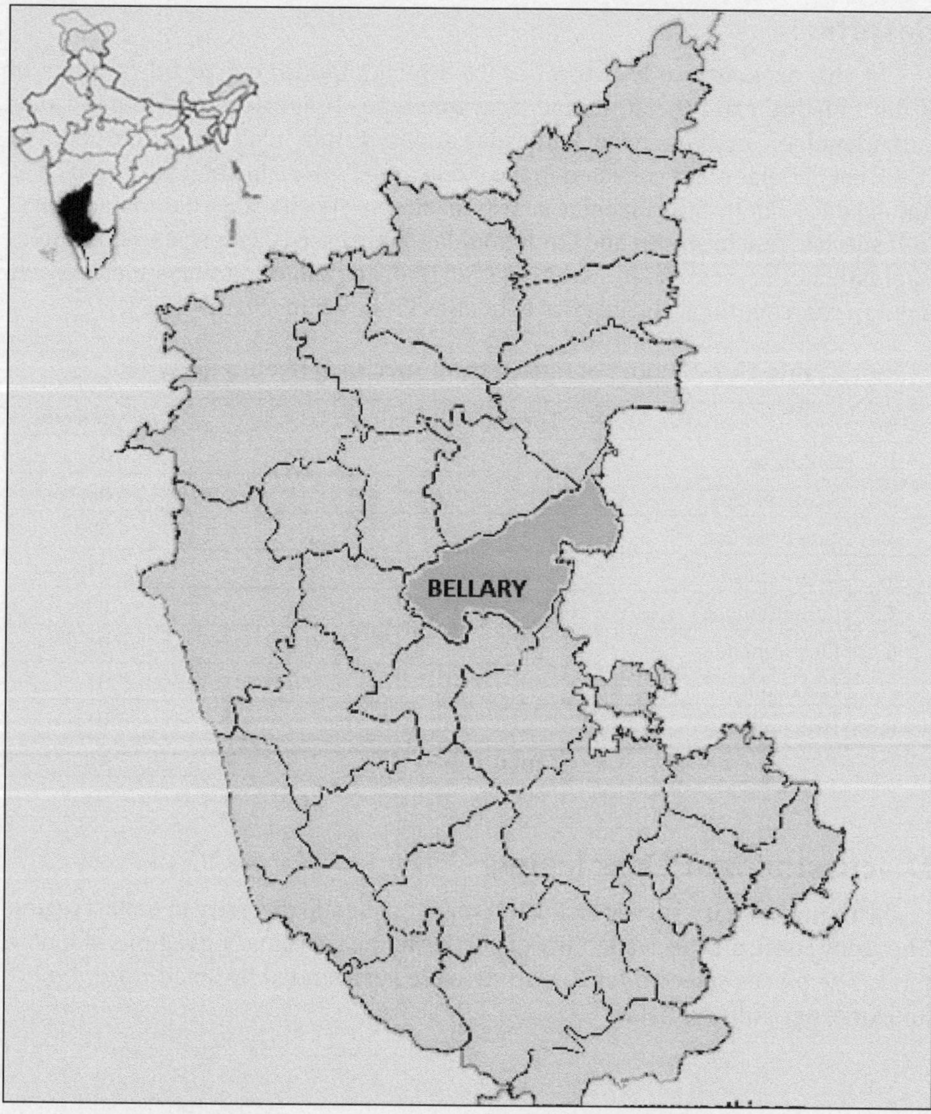

Figure 13.1: Geographical Location of Study Area.

Collection and Identification

The present study was carried out for six months (August 2019 to January 2020). The sampling of beetles was done every day by applying three standard methods such as net sweeping, hand picking and pitfall traps. After collection the process of identification was done by direct observation of beetles. The specimens identified were left in the field after identification as per Wildlife Protection Act 1972. The collected beetles were identified using the standard identification keys (Jonathan *et al.*, 1906., T. M. Ghosh, 1996, J. A. Russell, 2009) and bug guide net which is an open-source platform on web.

Results

In the present study a total of 16 species belonging to 08 families of Order coleopteran *viz.* Carabidae, Scarabaeidae, Tenebrionidae, Buprestidae, Carculionidae, Chrysomelidae, Elateridae and Hydrophylidae were recorded. The check list of Coleoptera collected in the study area is given in Tables 13.1 and 132. Among these family Scarabaeidae shares maximum species (05), Carabidae shares (03) species, Tenebronidae and Curculionidae share each (02) species. The families Buprestidae, Chrysomelidae, Elateridae and Hydrophilidae had single species each. A graph representing the diversity of beetles is shown in (Figure 13.2).

Table 13.1: Number of Families and Species of Beetles Recorded

Sl.No.	*Family*	*No of Species*
1	Carabidae	03
2	Scarabaeidae	05
3	Tenebronidae	02
4	Buprestidae	01
5	Curculionidae	02
6	Chrysomelidae	01
7	Elateridae	01
8	Hydrophilidae	01
	Total	**16**

Discussion and Conclusion

This preliminary survey is the first report of beetle diversity in Ballari region. The study confirms the availability of beetles in the surrounding regions of Ballari taluk. The places selected for this study were surrounded by small water bodies and large agricultural fields.

Table 13.2: Check List of Beetles in the Study Area

Sl.No.	*Family*	*Commonly known as*	*Genus and Species*
1	Carabidae	Ground Beetles	*Anthiasex guttata*
			Pheropsophus jessoensia
			Onthophagus coenobita
2	Scarabaeidae	Dung Beetles	*Oryctes rhinoceros*
			Heteronychus arator
			Scaptodera rhadamistus
			Anomala pallida
			Onthophagus gazella
3	Tenebronidae	Darkling Beetles	*Tenebrico obscurus*
			Tribolium destructor
4	Buprestidae	Metallic Wood-boring Beetles	*Eurythrea aurata*
5	Curculionidae	Snout Beetles	*Larinus planus*
			Myllocerus discolor
6	Chrysomelidae	Leaf Beetles	*Aulacophora foveicollis*
7	Elateridae	Click beetles	*Elater abruptus*
8	Hydrophilidae	Water Scavenger Beetles	*Hydrophilus piceus*

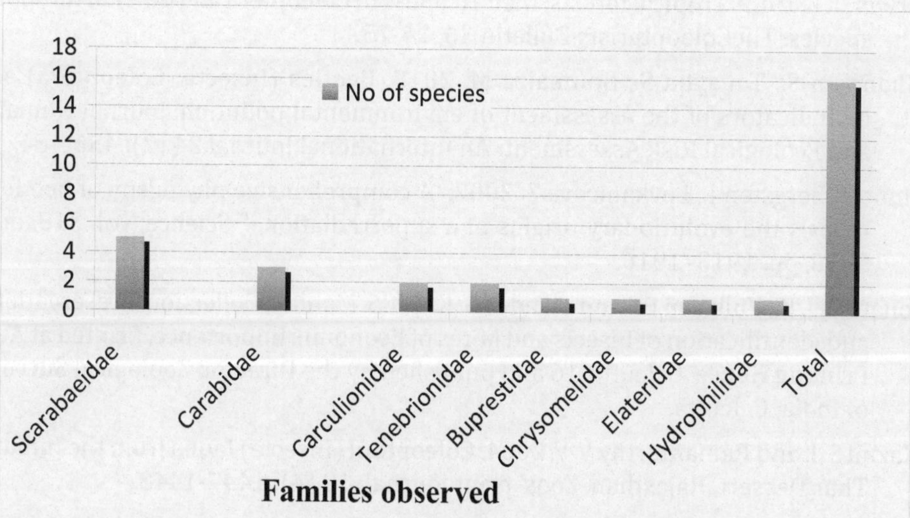

Figure 13.2: Family-Wise Distribution of Beetles in the Study Area.

These places can fulfil the food requirements as well provide the favourable conditions for beetles. Beetles play a very important role as a major part of the food-chain in invertebrates. This study will provide the basic data for researchers who have special interests in the conservation of beetles in this region. Further studies are required to be carried out in order to understand the composition, diversity index, species occurrence in different seasons and record other families as well in different regions of Ballari.

REFERENCES

Alfred J R B, Das A K, Sanyal A K. 1998. Faunal Diversity of India, ENVIS Centre, Zoological Survey of India, Calcutta. 259-268.

Choate P. M. 2001. Manual for identification of the ground beetles (Coleoptera: carabidae) (including tiger beetles) of Florida.

Choate P.M.1999. "Introduction to the Identification of Beetles (Coleoptera). Dichotomous Keys to Some Families of Florida Coleoptera, " pp. 23–33.

Crowson, R. A.1981. The biology of the Coleoptera. Academic Press: New York, xii + 802 pp.

Dabhade D.S., Shinde A.H., Tayade S.N., Kulkarni M.D., and Lohiya V.N. 2012. A study on beetle diversity in Mangrulpir Tahsil, Dist. Washim, Maharashtra. Multilogic in Science. Vol II, Issue III: 45-49.

Elzinga, A. 1992. "Political and Epistemic Regimes", in T. Brante, S. Fuller and W. Lynch (eds) Controversial Science, Albany, NY: SUNY Press. pp. 127-152.

Erwin, T. L. 1982. Tropical forests: their richness in Coleoptera and other arthropod species. TheColeopterists Bulletin 36: 74-75.

Ghannem S., Touaylia S., Boumaiza M. 2018. Beetles (Insecta: Coleoptera) as bioindicators of the assessment of environmental pollution. Journal Human and Ecological Risk Assessment: An International Journal.24 (2): 456-464.

Hunt T., Bergsten J., Levkanicova Z. 2007. "A comprehensive phylogeny of beetles reveals the evolutionary origins of a superradiation, " Science, vol. 318, no. 5858, pp. 1913–1916.

Jonathan J. K., Kulkarni P.P and Tikader B. K. 1986. Manual: Collection, Preservation and identification of Insects and Mites of Economic Importance. Printed at Art Printing House, Calcutta-16 and published by the Director, Zoological Survey of India, Calcutta.

Kazmi S. I. and Ramamurthy V V 2004. Coleoptera (Insecta) fauna from the Indian Thar Dessert, Rajasthan. Zoos' print journal, 19 (4): 1447-1448.

May, R. M. 1988. How many species are there on Earth? Science 241: 1441-1449.

Moitreyee Banerjee.2014.Diversity and Composition of Beetles (Order: Coleoptera) of Durgapur, West Bengal, India. Psyche A Journal of Entomology (9): 1-6.

Powell, J. A. 2009. Coleoptera, " in Encyclopedia of Insects, H. Vincent Resh and T. Ring Carde, Eds., Academic Press, New York, NY, USA, 2nd edition. 199 p.

Rosenzweig M.L.1995. Species Diversity in Space and Time, Cambridge University Press.

Satheesha D, Nikhath A.J, Vrushali M.M, Jayaraj F.C. and Sreenivasa G.2018. Preliminary study on composition and diversity of beetles (Order- Coleoptera) in and around Davangere University Campus, Davangere, Karnataka. Journal of Entomology and Zoology Studies. 6(4): 1751-1758.

ShirbhateMilind and Shirbhate Amrita.2020.Diversity and checklist of Beetles (Arthropoda: Coleoptera) from Forest areas and Agricultural areas of District Akola, (Maharashtra), India. Environment Conservation Journal, 21 (1 and 2): 89-94.

UniyalV. P. 2001. Conserving biodiversity in the Indian Trans- Himalaya: New initiatives of field conservation in Ladakh. Annual technical report submitted to wildlife institute of India-Dehradun.

Wilson, E. O. 1992. The diversity of life. Allen Lane The Penguin Press, Massachussetts.

Chapter 14

Comparative Study of Bacterial and Fungal Diversity in Paddy and Ginger Fields of Chikkamagluru Region, Karnataka

D.S. Somashekar[1], Shashikanth H. Majagi[2] and Dhananjaya S.G.[3]

[1]PG Studies in Zoology, I.D.S.G. Government College, Chikkamagluur – 577 102, Karnataka
[2]Department of Studies in Zoology, VSK University, Ballari, Karnataka
[3]Govt. Science College, Chitradurga, Karnataka
e-mail: [1]drsomashekar7@gmail.com; [2]smajgi@rediffmail.com

The diversity and composition of soil microbe communities is thought to have a direct influence on wide range of ecosystem process. Microorganisms are perhaps the most diverse and abundant, which exceeds that of eukaryotic organisms on earth. It is widely accepted that the extent of microbial diversity has not been adequately explored. In the present study soil samples were collected from Paddy and Ginger fields in the vicinity of Chikkamagluru region. A total of 12 distinct strains of bacteria were recorded such as Actionomycetes, Bacillus subtilis, Pseudomonas aeruginosa, Aeromonas, Proteus species, Clostridium species and Acinetobacter species. Twelve distinct types of fungi were recorded in the selected agro ecosystems with Sordariomycetes species to be the dominant, the least generas were Mucor species. Out of 24 species of Bacteria and fungi, 13 species are found both in paddy and Ginger field, 8 species recorded only in paddy field and 3 species were found only in Ginger field.

***Keywords**: Chikkamagluru, Microorganisms, Paddy, Ginger, Bacillus subtilis.*

Introduction

Microorganisms are very diverse and include all bacteria, archea and fungi, algae protozoa. Microorganisms live in every part of the biosphere, including soil, hot springs inside rocks. Single celled, prokaryotic microorganisms were first farm of life to develop on earth, approximately 4 billion years ago, most of the microbes reproduce rapidly and in great number. Microbes in the soil are directly tied to nutrient recycling especially carbon, nitrogen and phosphorus and sulfur bacteria are a major class of microorganisms that keep soils healthy and productive. The gaseous nitrogen of the atmosphere represents a vast store of potential fertility. It is not directly available to plants. Nitrogen-fixing bacteria, how-ever, absorb this gas from the soil solution and convert it to cell protein. When the cells die, other microbes attack the protein and convert the nitrogenous constituents to ammonium, which then becomes an available nutrient. Similarly, microbial action on plant and animal residues releases the many combined nutrients from their unavailable forms. In many instances microbial actions results in indirect as well as direct production of active fertility.

General microbial activity in soil it require certain abiotic factors in a moderate condition such as moisture, temperature, aeration, pH, food supplements, *etc.* Among the different microorganisms inhabiting in the soil, bacteria are the most abundant and predominant organisms. Most fungi are inconspicuous because of the small size of their structure, and their cryptic lifestyles in soil or on dead matter fungi includes symbionts of plants, animals or other fungi and also parasites, they may become noticeable when fruiting, either as mushrooms or as moulds. Fungi perform an essential role in the decomposition of organic matter and have fundamental roles in nutrient cycling and exchange in the environment.

Chhnokar and Subberrao (1967) isolated fungal cultures from legume root nodules and identified as *Aspergillus flavus*, *Aspergillus niger* and *Aspergillus nidulans* based on their spore structure colony morphology. Gowsalya *et al.* (2000) they worked on isolation of bacteria from soil sample for exopolysaccharide production. For bacterial isolation the soil samples were collected from farmland at Gandarvakottai, Pudukkottai district, microorganisms were separated by serial dilution method. Each of the isolates were experimented for the morphological characteristic like shape, gram nature and arrangement of cells, motility *etc.* Rakeshkumar *et al.* (2004) worked on isolation, screening and identification of fungi from soil of Bhopal at some different locations. Matsuyama *et al.* (2007); Elizabeth Pradel *et al.* (2007) and Sungano *et al.* (2005) studied that bacterial community in plant residues in a Japanese paddy field using DGGE and T-RFLP they found that members of firmicutes (clostridia). Alpha, Gamma, Delta- Proteobacteria, Nitrospira, Acidobacteria and Spirochaetes were the predominant micro organisms in the rice residues.

Kostadinov *et al.* (2006) investigated isolation and identification of filamentous fungi from soil, among 12 belonging to the phylum *Ascomycota* (7 genera),

Deuteromycota (2genera), *zygomycota* (2genera) and *Basidiomycota* (1genera), *Mucorcladosporium*, Alternaria, *Aspergillus* and *Pencillium* were predominant genera. *Locanicillium, Batrytis, Geomyces, Monodictys* and *Rhyzopus* were the most frequently isolated genera. Most of the fungal isolates proved to be cold tolerant. Brown *et al.* (2005) studied the Macrofungal diversity in fragmented and disturbed forests of the Western Ghats of India.

The present investigation was carried out to know the diversity and distribution of fungal and bacterial species in two agro ecosystems *viz.*, Paddy and Ginger fields at Chikkamagluru vicinity of Karnataka state.

Study Area

Chikkamagaluru is situated roughly in the south western part of Karnataka state. A large are at this districts in Malnadi'e, forested hilly region of heavy rain fall. Chikkamagaluru is located between 13°15' 53" north latitude and between 75° 04'46" and 76°21'50" east latitude. Its greatest length from east to west is 138.4 km and from north to south is 88.5km. It is located at an elevation of 3400 feet. The temperature ranges between 11 – 20°c during winter and 25-34°c during summer. The annual precipitation is 2000-2540 mm and considerably higher than the 750-1000 mm.

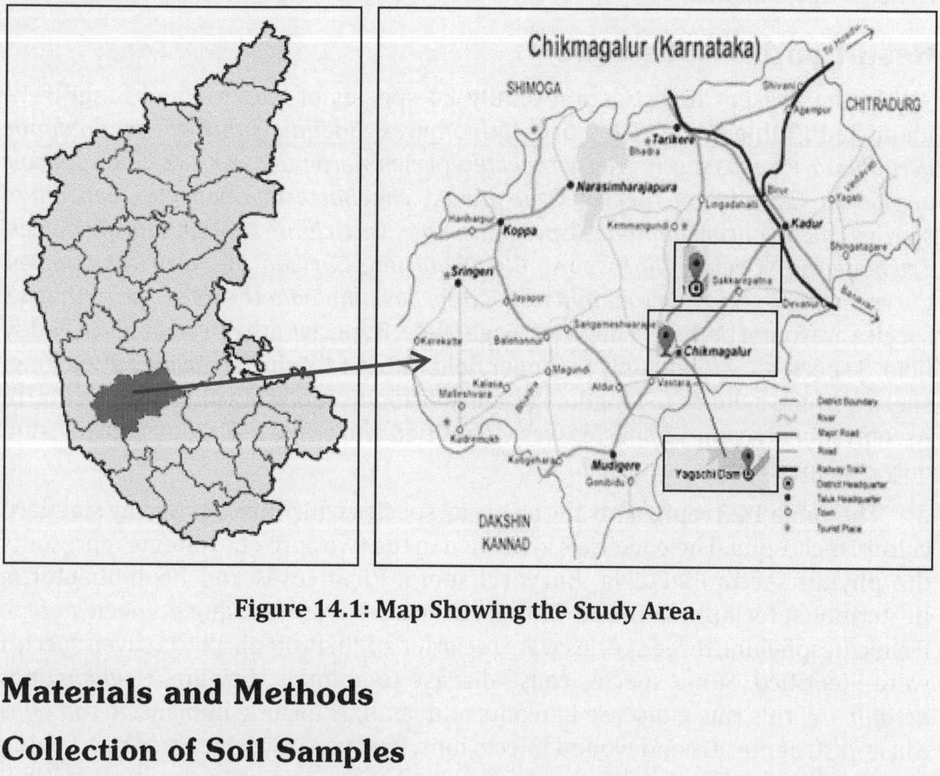

Figure 14.1: Map Showing the Study Area.

Materials and Methods

Collection of Soil Samples

Soil samples are collected from paddy fields and ginger fields of Chikkamagluru region. It was taken by means of sterilized spatula by digging. The soil sample to a depth of 10cm and collected in sterile polythene bags, the samples were then brought to the laboratory for microbiological study.

Culturing of Bacteria

Nutrient agar media was used for culture of bacteria, nutrient agar is a general purpose medium which supporting growth of wide range of non-fastidious organisms. 1gm of soil in a100ml of water mixed thoroughly, then by using streaking plate method that sample inoculum is streaked on agar media, which is placed in Laminar air flow by using inoculation loop.

Culturing of Fungi

Potato dextrose agar is the most widely used medium for growth of fungi and bacteria. Potato Dextrose agar is composed of dehydrated potato infusion and Dextrose that encourage luxuriant fungal growth. Agar is added as the solidifying agent. Many standard procedures use a specified amount of sterile tartaric acid (10 per cent) to lower the pH of this medium to 3.5±0.1 inhibiting bacterial growth.

These colonies are isolated and perform pure culture, by using streak plate technique isolate single kind of organisms. Bacteria and Fungi we are identified as per Bergy's manual.

Result and Discussions

In the present investigation totally 24 species of bacteria and fungi were identified (Table 14.1). They are *Actinomyces, Bacillus subtilis, Pseudomonas aeruginosa, Proteus species, Acenetobacter*species, *Aeromonas*species, *Enterobacter aerogenes, Clostridium* species, *Oxobacter, Archaebacteria, Geobacter, Pencillium* species, *Microsporum gallinae, Aspergillus niger, Verticillum* species, *Mucor* species, *Trichoderma* species, *Nigrospora, Cladosporium, Curvularia, Fusarium* species, *Chrysomonas* species, *Acromonium strictum, Lasiodiplodia theobromae*. In that 13 species are found both in Paddy and Ginger field, 8 species are present only in Paddy field, 3 species are found only in Ginger field. Table 14.2 shows the identified fungi species which belongs to kingdom fungi, division Ascomycota and Zygomycota. In Ascomycota division 11 species were identified and while in Zygomycota division only one species was recorded.

The Table 14.3 represents the bacterial species which are isolated by standard culture technique. The bacteria's identified in the two agro ecosystems represents the phylum Actinobacteria, Euryarchaeota, Firmicutes and Proteobacteria. In Actinobacteriaphylum and Euryarchaeota represents single species each. Firmicutesphylum, three species were recorder andin Proteobacteria seven species were identified. Some species cause disease to animals they are, *Pseudomonas aeruginosa*, this cause disease in plants and animals include humans. *Aeromonas* cause gastroenteritis and wound infects ions. In severe form it causes necrotizing fasciitis disease. Then *Enterobacter* species cause respiratory, gastrointestinal

Table 14.1: Distribution of Bacteria and Fungi Species in a Study Area

Sl.No.	Species name	Paddy field	Ginger field
1	*Actinomyces*	+	+
2	*Bacillus subtilis*	+	+
3	*Pseudomonas aeruginosa*	+	+
4	*Proteus*	+	–
5	*Acinetobacter* species	+	+
6	*Aeromonas* species	+	–
7	*Enterobacteraero genes*	+	+
8	*Clostridium* species	+	+
9	*Oxobacter* species	+	–
1	*Archae* bacteria	+	–
11	*Geobacter* species	+	–
12	*Pencilium* species	+	+
13	*Microsporum gallinae*	+	–
14	*Aspergillus niger*	+	+
15	*Verticillium* species	+	+
16	*Mucor* species	+	+
17	*Trichoderma*	+	–
18	*Nigrospora*	+	+
19	*Cladosparium*	+	–
2	*Fusarium* species	+	+
21	*Curvularia* species	+	+
22	*Chrysomonas* bacteria	–	+
23	*Acromonium strictum*	–	+
24	*Lasiodiplodia theobromae*	–	+

+: Presence and –: Absence.

and urinary tract infections. *Clastrodium* species is a significant pathogen it's a important cause for diarrhea. *Fusarium* species, it causes opportunistic infections in humans cause abdominal pain, diarrohea, vomiting *etc.*, Variations in the growth of microbes were noticed in the fields because indiscriminate use of variety of pesticides in the crop fields.

Microbes in the soil are directly tied to nutrient recycling especially carbon, nitrogen and phosphorus and sulfur bacteria are a major class of microorganisms that keep soils healthy and productive. The gaseous nitrogen of the atmosphere represents a vast store of potential fertility. It is not directly available to plants. Nitrogen-fixing bacteria, how-ever, absorb this gas from the soil solution and convert it to cell protein. When the cells die, other microbes attack the protein

Table 14.2: List of Fungi Species Identified in Study Area

Division	***Class***	***Species Names***
Ascomycota	Dothideomycetes	*Cladosparium* species
	Euascomycetes	*Curvularia lunata*
	Eurotiomycetes	*Aspergillus niger*
		Pencillium species
		[illegible]
	Incertaesedis	*Verticillium* species
	Sordariomycetes	*Trichoderma* species
		Nigrospora sphaerica
		Fusarium species
		Acromonium strictum
		Lasiodiplodia theobromae
Zygomycota	Mucormycotina	*Mucor* species

Table 14.3: List of Bacteria Species Identified in the Study Area

Phylum	***Class***	***Order***	***Family***	***Species Names***
Actinobacteria	Actinobacteria	Actinomycetales	Actino-mycetaceae	*Actinomyces* species
Euryarchaeota	Methano-bacteria	Methano-bacteriales	Methano-bacteriaceae	*Archae* bacteria
Firmicutes	Clostridia	Clostridiales	Peptostrep-tococcaceae	*Clostridium* species
			Clostridiaceae	*Oxobacter* species
	Bacilli	Bacillales	Bacillaceae	*Bacillus subtilis*
Proteobacteria	Delta Proteobacteria	Desulfuro-monadales	Geobacteraceae	*Geobacter* species
	Gamma Proteobacteria	Aeromonadales	Aeromonadaceae	*Aeromonas* species
		Entero-bacteriales	Entero-bacteriaceae	*Enterobacter* species
				Proteus species
		Pseudo-monadales	Pseudo-monadaceae	*Chrysomonas* species
				Pseudomonas aeruginosa
			Moraxellaceae	*Acinetobacter* species

and convert the nitrogenous constituents to ammonium, which then becomes an available nutrient. Similarly, microbial action on plant and animal residues releases the many combined nutrients from their unavailable forms. In many instances microbial action results in indirect as well as direct production of active fertility.

Benilasmily, *et al.* (1979) they reported on soil micro flora of paddy field among different rice forming system at Kumarakum region, south west part of kuttanadu Kerala from two paddy fields each with single rice forming and fish rice rotational forming A total of 11 bacterial species were isolated from paddy field soil where as 15 were obtained from fish rice rotational forming soil the highest frequency of occurrence among paddy field isolates was the bacillus species followed by *klebsiella pneumonia, k. oxytoca* and pseudomonas species similarly the fungal species in rice field with paddy alone showed 13 species and fish rotational field with paddy showed 19 fungal species *Aspergillus* species recorded highest frequency of occurrence among the paddy field isolates. Johnson (1954) isolated fungal cultures from rhizophere soil and identified as *Aspergillus niger, Aspergillus terreus* and *Pencillium digitatum* based on their spore structure and colony morphology. Singh *et al.* (1984) identified phosphate solubalising fungal cultures *viz. Aspergillus awamori* and *Aspergillus niger* based on their colony colour, colony morphology and spore structure.

Ploetz.*et al.* (1985) studied, soil sample to assayed for plant pathogenic andnon pathogenic fungi. They recorde16genera of fungi identified species of *Aspergilus, Fusarium, Pencillium, Rhizopus* and *Trichoderma* recorded for upto 75 per cent of the total fungal population detected. Kuk-jeong chin, *et al.* (1998) worked on characterization and identification of numerically abundant culturable bacteria from the anoxic bulk soil of Rice paddy microcosm. Most probable number (liquid serial dilution culture) counts were obtained for polysaccherolytic and saccherolytic fermenting bacteria in the anoxic bulk soil of flooded microcosms containing rice plants. Gowsalya *et al.* (2000) worked on isolation of bacteria from soil sample for exopolysaccharide production. Similar observations were recorded by different workers *viz.*, Rakeshkumar *et al.* (2004); Kostadinov *et al.* (2006); and Rahman *et al.* (2020)

REFERENCES

Bergey's Manual® of Systematic Bacteriology: Volume 2: The Proteobacteria, Part B: The Gammaproteobacteria Volume 2, Page 2 of Bergey's Manual of Systematic Bacteriology.

Brown, N., Bhagwat, S. and Watkinson, S. (2006), Macrofungal diversity in fragmented and disturbed forests of the Western Ghats of India. Journal of Applied Ecology, 43: 11-17.

Chhonkar P K And N. S. Subba-Rao (1967) Phosphate solubilization by fungi associated with legume root nodules. Canadian Journal of Microbiology Volume 13, Number 7

Elizabeth Pradel, Yun Zhang, Nathalie Pujol, Tohey Matsuyama, Cornelia Bargmann, Jonathan J. Ewbank 2007 Detection and avoidance of a natural product from the pathogenic bacterium *Serratia marcescens* by *Caenorhab ditiselegans* Proceedings of the National Academy of Sciences Feb 2007, 104 (7) 2295-2300

Gowsalya, A., Ponnusami, V. and Sugumaran, K. (2014) Isolation of Bacteria from Soil Sample for Exo-Polysaccharide Production. International Journal of ChemTech Research, 6, 2925-2928.

Johnson, 1954, "Isolated fungal culture from rhizophore soil". V-4. P. 58-70.

Kostadinov S, Zlatic M, Dragovic N, Gavrilovic Z. (2006): Soil Erosion in Serbia and Montenegro, Soil Erosion in Europe, p.p. 271-277; Editors: John Bordman and Jean Poesen; John Wiley and Sons, Ltd; London

Kuk-Jeong Chin, Frederick A. Rainey, Peter H. Janssen, Ralf Conrad, (1998) Methanogenic Degradation of Polysaccharides and the Characterization of Polysaccharolytic Clostridia from Anoxic Rice Field Soil, Systematic and Applied Microbiology, Volume 21, Issue 2,

Matsuyama T, Nakajima Y, Matsuya K, Ikenaga M, Asakawa S, Kimura M (2007) Bacterial community in plant residues in a Japanese paddy field estimated by RFLP and DGGE analyses. Soil Biology and Biochemistry 39 (2): 463-472.

Ploetz R C, Mitchell D J and Gallaher R N 1985 Population dy-namics of soil-borne fungi in a field multi-cropped to rye and soybeans under reduced tillage in Florida.Pythopathology 75, 1447–1451.

Rahman, M.M., Lisa, L.A., Ahmed, S., Uddin, M.N., Hasan, T. and Zaman, S. (2020) Isolation and Characterization of Beneficial Microorganisms in Organic, Semi-Organic and Conventional Fertilizer Treated Agricultural Field Soil and Comparison of Bacterial Richness. *Journal of Agricultural Chemistry and Environment*, 9, 223-239

RakeshKumar R. Kirti Saurabh Narendra Kumawat Prem K. Sundaram Janki Sharan Mishra Dhiraj K. Singh Hansraj Hans Bal Krishna Bhagwati Prasad Bhatt (2021) Sustaining Productivity Through Integrated Use of Microbes in Agriculture. In: Seneviratne G., Zavahir J.S. (eds) Role of Microbial Communities for Sustainability. Microorganisms for Sustainability, vol 29.Springer, Singapore.

Singh *et al.*, 1984", Identification of phosphate solubilizing fungal culture". Modern soil microbiology.fifth edition. P-1100-1123.

Sugano A, Tsuchimoto H, Tun CC, Asakawa S, Kimura M (2005) Succession and phylogenetic profile of eubacterial communities in rice straw incorporated into a rice field: Estimation by PCR-DGGE Analysis. Soil Science and Plant Nutrition 51 (1): 51-60.

Tomoyuki Hori *et al.*, 2009, Iron reducing microbes in paddy soil. V-16. P: 567-786.

Chapter 15

Catfish Biodiversity of Tunga and Bhadra Rivers of Western Ghats, Karnataka, India

*Shahnawaz Ahmad[1] *, M. Venkateshwaralu[2], Farooq Ahmad Ganai[1] and Farooq Ahmad Mir[1]*

[1]P.G. Department of Zoology, Govt. Degree College Sopore, Kashmir
[2]Department of Applied Zoology, Bio-Science Block, Kuvempu University, Shimoga – 577 451, Karnataka
**Corresponding author e-mail: snahmad78@gmail.com*

An extensive ichthyological study was carried from randomly selected 14 sites of Tunga and Bhadra rivers, Western Ghats, Karnataka. A total number of 20 catfish species were recorded belonging to 8 families. Out of 8 families, catfish occurrence was dominated by Bagridae, Siluridae, Sisoridae, Schilbeidae, Channidae, Notopteridae, Claridae and Heteropneustidae in decreasing order respectively. This research paper provides the checklist of catfishes which would help in formulating conservation planning for these rivers as catfishes are coming under threatening categories by overfishing, dynamiting, pollution threats and introduction of exotic fishes.

***Keywords**: Catfishes, Biodiversity, Tunga and Bhadra rivers, Western Ghats.*

Introduction

A catfish belong to an order known as siluriformes (Teleostei) and is a diverse group of fishes that represent more than 3, 000 species, 478 genera and 36 families,

Ferraris, and Pinna, (1999). They constitute 1/3 of the freshwater fish biodiversity of the world. In Indian subcontinent, they comprise about 158 species of inland catfishes from 50 genera and 13 families. The Indian catfish includes 13 families, bagridae, siluridae, schilbeidae, pangasiidae, amblycipitidae, sisoridae, clariidae, heteropneustidae, chacidae, olyridae, akysidae, ariidae and plotosidae, Talwar, and Jhingran, (1991). The diversity of catfish in India is the highest in areas like Western Ghats, Northeastern hills, Gangetic river system, Barman, (1994); Thomas *et al.* (2002). The diverse habitats of catfishes range from upland streams, large river channels and seasonal flood to plain lagoons and prefer broad feeding niches like detritivores, benthic insectivores and piscivores, Winemiller, and Winemiller, (1996).

The Western Ghats is a 1600 km long, unbroken chain of mountains along the west coast of Peninsular India, which is recognized as a global biodiversity hot spot (Mittermeir *et al.*, 2000; Bossuyt, 2004). It is a region with variety of vegetation types and climatic zones that supports remarkable endemism of variety of animals including fish fauna (Bhat and Jayaram, 2003). The Western Ghats is so called because of its occupying position in Peninsular India between 8 to 40° N and 7 to 77° E from the Tapti Valley in Gujarat to Kanyakumari in Tamil Nadu. These series of hill ranges run north south along the west coast traveling the states of Gujarat, Maharashtra, Goa, Karnataka, Kerala and Tamil Nadu.

The Karnataka part of Western Ghats lies between the latitude of 11° 12' N and 18° 12' E, which runs north to south for about 640 km with an average width of 50-65 km. The Karnataka state is rich in water resources with 300 square km of coastline, 27,000 square km of continental shelf, 2.93 lakh hectare area under tanks, 2.27 lakh hectare reservoirs, 8,000 hectare of estuaries and 5,813 square km of rivers. The Karnataka is endowed with several riverine systems, among them river Tunga and Bhadra have been selected for the present study. In the present paper, I will document the families of catfishes in these two selected rives based on their diversity, range, and basic ecology. The diversity of catfish fishes will follow the recent phylogeny of Sullivan *et al.* (2006) with additions by Lundberg *et al.* (2007). Although other phylogenies for catfishes are available (Hardman 2005). Teugels, (2003) reviews the characters that define catfish families and major groups, and Diogo, (2003) reviews phylogenetic relationships.

Hence, the aim of this present paper is the documentation of catfish diversity in Tunga and Bhadra rivers which would be useful in conservation planning

Materials and Methods

Gangamula, the source of both the Tunga and Bhadra rivers, is situated at 13° 15' N Latitude and 75° 14' E Longitude. The hill from which they originate is called the Varaha Paravatha, that confines to the Koppa and Balehonnur taluks of Karnataka. The Tunga and Bhadra rivers are the major sub-basins of the east flowing largest peninsular river, the Krishna and two west flowing rivers that are Seetanadi and Netravati. The course of Tunga and Bhadra rivers are obstructed at several places for human use like hydroelectric projects and the main purpose of this is irrigation. The Tunga river is dammed for irrigation and fishery purposes at Gajanur place about 15 kms from Shimoga. The Bharda river has a big reservoir project near Lakkavalli for power irrigation. Upstream of Bhadra river are inside the Bharda Reserve Forest and that is why they are less disturbed by human activities, whereas the main Tunga river passes through many habitations, mostly pilgrim sites.

The 130 km river Tunga, which flows for about 64.4 km in Shimoga district with catchment area of 22,39.4 square km. Its course is at first north-east passes Shringere town to Baggunji, where it turns north-west direction and then continues up to Thirthahalli. It is a medium, perennial river and for most of the year, the riverbed flows thinly. The water is abundant during the period of June to October and becomes a good source for drinking and irrigation purposes. Its banks are not steep like that of Bhadra river and its streams are generally considered as more healthy and less disturbed.

The Bhadra river rises close to the Tunga in the Gangamula in the Varaha Hills of Western Ghats, 15 miles west of Kalasa with fast flow of water. It flows in an eastern direction and then joins at Sangameshwara by the Anebidda halla from the south, then turns north-east and runs past Khandeya forest, across the mouth of the Jagar valley. The drainages of which it receives in the Somavahini at Hebbe and then nearby Gakvalli by small unnamed streams from Kalhatti, Giri and Kalgurga join with the main river and the course continues upto Benkipur and Hollehonnur until the point of confluence with Junga at Kudli in the Shimoga district. The river Bhadra is fairly a big river and flows almost throughout the year. Seven (7) study sites each from Tunga and Bhadra rivers, were selected for the present study (Figure 15.1) and the general geographical features of selected research areas or sites is summarised in Table 15.1. However, fishes were collected by using monofilament and multifilament gill nets of various mesh sizes ranging from 6-15 mm, drag nets, scoop nets and cast nets. Fishes were examined, counted and then were released back into the system after taking few specimens (5-10) which were preserved in buffered formalin (10 per cent) and transported carefully to the laboratory for further analysis and were deposited at the department of Applied Zoology, kuvempu University. Fishes were identified based on keys for Fishes of the Indian Subcontinent (Jayaram 1999; Talwar and Jhingran 1991; Dutta Munshi and Srivastava, 2006).

Results and Discussion

A total of twenty catfish species were recorded and examined belonging to eight families with 18.18 per cent from 14 different randomly selected sites of Tunga and Bhadra rivers, Western Ghats. It is evident from the present study that catfish family namely Bagridae was more diversified and predominant than other recorded families like Siluridae, Sisoridae, Schilbeidae, Channidae, Claridae, Heteropneustidae and Notopteridae respectively in Tunga than Bhadra rivers

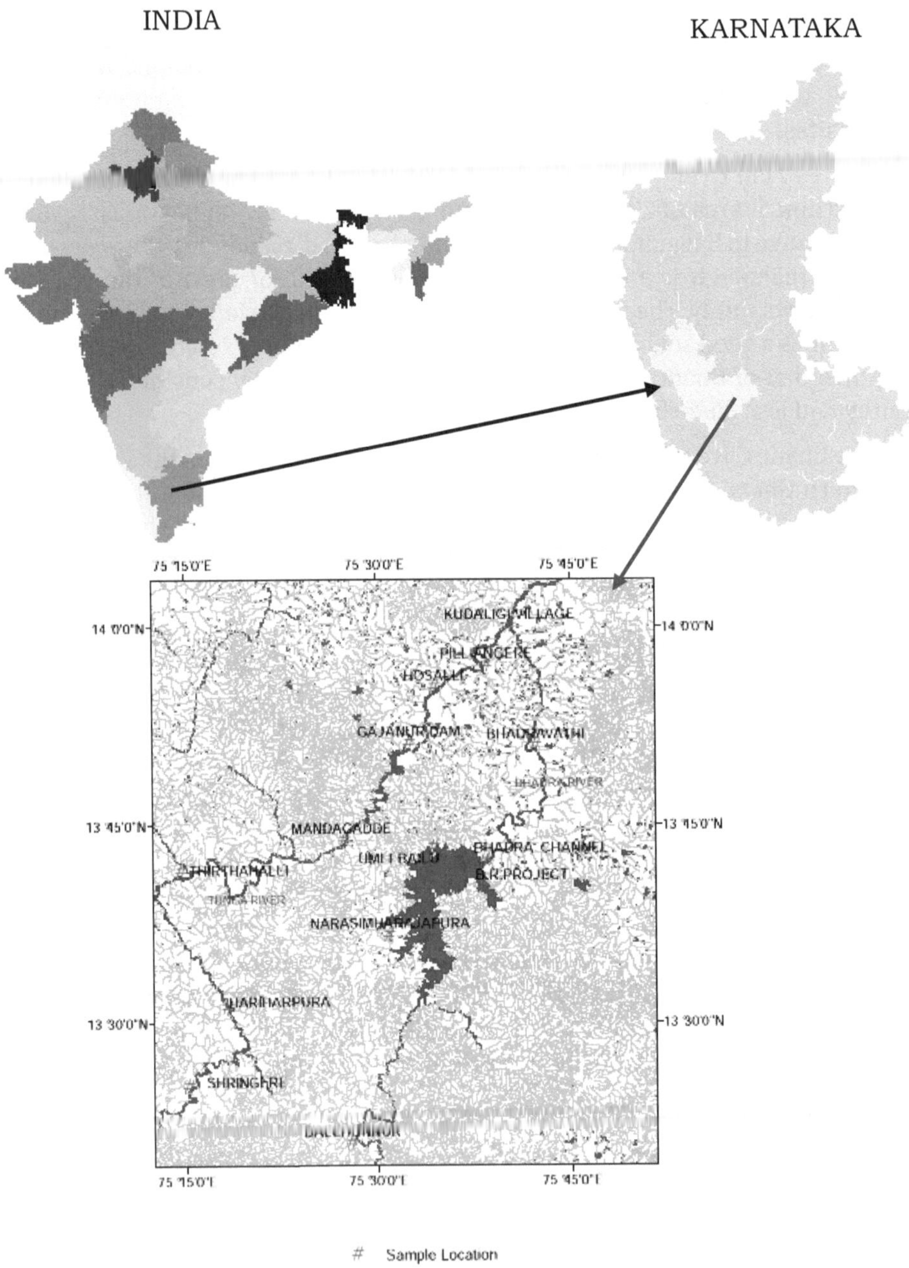

Figure 15.1: Location of the Study Area

(Table 15.2). It was also recorded that catfishes were mostly found in lower and deep reaches of the selected rivers. Moreover, the range of distribution would be because of the fact that Tunga river is more pristine and protected in the form of fish sanctuaries and undisturbed areas than Bhadra river which is much disturbed in the form of pollution threats and excess fish catch at various locations like B. R. Project, N.R. Pura and Bhadravathi, (Shahnawaz *et al.*, 2010). Relatively high-protected areas are characterized by diverse fish species in a variety of habitats (Arunachalam, 2000). The study is supported by Kar and Sen, 2007 who reported the same observation in their studies.

Table 15.1: General Geographical Features of Tunga and Bhadra Rivers, Western Ghats

Sl.No.	*Location*	*Altitude (m)*	*Latitude*	*Longitude*	*Forest Type*
Tunga river					
1	Shringere	680	13º, 56'N	75º, 84'E	Dry deciduous
2	Hariharpura	620	13º, 31'N	75º, 18'E	Dry deciduous
3	Thirthahalli	625	13º, 41'N	75º, 14'E	Dry deciduous
4	Mandgadde	615	13º, 44'N	75º, 27'E	Agricultural
5	Gajanur Dam	580	13º, 50'N	75º, 31'E	Dry deciduous
6	Hosalli	570	13º, 52'N	75º, 34'E	Agricultural
7	Pilliangere	550	13º, 57'N	75º, 38'E	Agricultural
Bhadra river					
1	Balehonnur	766	13º, 21'N	75º, 26'E	Dry deciduous
2	N.R. Pura	660	13º, 37'N	75º, 32'E	Dry deciduous
3	Umlebailu	665	13º, 43'N	75º, 34'E	Agricultural
4	B. R. Project	700	13º, 42'N	75º, 37'E	Dry deciduous
5	Bhadra channel	630	13º, 43'N	75º, 38'E	Agricultural
6	Bhadravathi	580	13º, 50'N	75º, 41'E	Dry deciduous
7	Kudaligi village	600	14º, 0'N	75º, 40'E	Dry deciduous

The present study shows that the species like *Wallago attu, Nangra itchkeea, Mystus krishnensis,* are in critical condition which needs protection of habitats. These species were represented only in one site and were not found in other sites and therefore, they were characterized by low abundance, narrow distribution and high threats. The decline is attributed to several factors including indiscriminate fishing, flow regimes, land-use practices, shrinkage, deforestation, construction of dams and barrages, introduction of exotic fish species *etc.* (Gadgil, 2004, Shahnawaz *et al.*, 2010).

Big sized catfishes including *Mystus cavasius, Sperata seenghala, Sperata oar* and *Clarias batrachus* are preferred as food fishes but are at moderate risk and this might be because of their diversified distribution in almost all the study sites.

Table 15.2: Shows the Catfish Biodiversity Richness Family-wise in Tunga and Bhadra Rivers, Western Ghats

Sl.No.	*Families/Species*	*Common Name (English)*
	Family I: Bagridae	
1	*Mystus cavasius* (Hamilton-Buchanan)	Gangetic mystus
2	*Mystus armatus* (Day)	Gangetic mystus
3	*Sperata (Aorichthys) aor* (Hamilton)	Whiskered catfish
4	*Sperata (Aorichthys) seenghala* (Sykes)	*Giant river catfish*
5	*Rita gogra* (Sykes)	Velvet catfish
6	*Rita pavimentatus* (Valenciennes)	Velvet catfish
7	*Mystus krishnensis* (Ramakrishniah)	Krisna mystus
8	*Mystus malabaricus* (Jerdon)	Jerdons mystus
	Family II: Siluridae	
9	*Ompok bimaculatus* (Bloch)	*Butter/glass catfish*
10	*Ompok pabo* (Hamilton-Buchanan)	*Two stripe gulper catfish*
11	*Wallago attu* (Schneider)	*Helicopter catfish*
	Family III: Claridae	
12	*Clarias batrachus* (Linn)	*Walking catfish*
	Family IV: Heteropneustidae	
13	*Heteropneustes fossilis* (Bloch)	*Stinging catfish*
	Family V: Sisoridae	
14	*Nangra itchkeea* (Sykes)	*Nangra*
15	*Glyptothorax lonah* (Sykes)	*Kanta/bandegirlu*
	Family VI: Schilbeidae	
16	*Proeutropicthys taakree* (Sykes)	*Indian taakree*
17	*Neotropius khavalchor* (Kulkarni)	*Khavalchor catfish*
	Family VII: Channidae	
18	*Channa punctatus* (Bloch)	*Green snakehead*
19	*Channa marulius* (Hamilton-Buchanan	*Bullseye snakehead*
	Family VIII: Notopteridae	
20	*Notopterus notopterus* (Hamilton)	*Bronze featherback*

We identified different factors including overfishing, hook fishing and dynamite fishing to be the most destructive threats to all fishes including catfishes also. These fishing practices have caused severe habitat destruction and declines of rare loach fishes and many important native food fishes. An example of this is *Wallago attu, Nangra itchkeea, Mystus krishnensis* which we observed have been facing extreme population decline across the landscape, mainly attributed to dynamite-fishing. Pollution, dumping of wastes from garbage and steel industry near Bhadrahalli site

(Bhadra River) and introduction of invasive exotic food fishes into dam reservoirs like B.R. Project Dam and lowland river sites were the other most critical threats to fishes in general but catfishes in particular. Kar and Sen, 2007; Karmakar and Das, 2007 recorded the same observation in their studies about the catfishes.

In the present study no introduced and exotic fish species has been witnessed in the upstream sites of both the Tunga and Bhadra rivers. That is why the introduction of exotic species to the native fauna is very essential factor when we think of conservation measures. Many upstreams are still pristine and free of exotic/introduced fish species which harbor great fish diversity. Therefore, they need great attention and where conservation efforts should be targeted.

REFERENCES

Arunachalam, M. 2000. Assemblage structure of stream fishes in the Western Ghats, (India). *Hydrobiologia.*, 430: 1-31.

Barman, R. P. 1994, "Fish fauna of Tripura, North-East India", *Journal of the Bombay Natural History Society.* 91, pp. 37-46.

Bhat, A. and Jayaram, K.C. 2003. A new species of the genus Batasio Blyth (Siluriformes: Bagridae) from Sharavathi River, Uttara Kannada, Karnataka. Zoos Print, 19(2), 1339-1342.

Bossuyt, F. 2004. Local endemism within the Western Ghats, Sri Lanka, biodiversity hotspots. *Science*, 306: 479-481.

Datta Munshi, J.S. and M.P. Srivastava. 2006. Natural history of fishes and systematics of freshwater fishes of India. Narendra Publishing House, Delhi, pp. 394.

Diogo, R. 2003. Higher-level phylogeny of Siluriformes: an overview. Pages 353–384 *in* G. Arratia, B. G. Kapoor, M. Chardon, and R. Diogo, editors. Catfishes, Volume 1. Science Publishers, Enfi eld, New Hampshire.

Ferraris, C. J. Jr. and de Pinna, M. C. C. 1999, "Higher-level Names for Catfishes (Actinopterygii: Ostariophysi: Siluriformes)". Proceedings of the California Academy of Sciences, 51, pp. 1–17

Gadgil, M. 2004. Karnataka State Biodiversity Strategy and Action Plan (KSBSAP), ENVIS, Technical Report No. 15, 418 pp. (http: //www.ces.iisc.ernet.in./hpg/envis).418pp.

Hardman, M. 2005. The phylogenetic relationships among non-diplomystid catfishes as inferred from mitochondrial cytochrome b sequences; the search for the ictalurid sister taxon (Otophysi: Siluriformes). Molecular Phylogenetics and Evolution, 37: 700–720.

Jayaram, K.C. 1999. The freshwater fishes of the Indian region. Narendra Publishing House, Delhi-06.

Kar, D. and Sen, N. (2007), "Systematic List and Distribution of Fishes in Mizoram, Tripura and Barak Drainage of North Eastern India", *Zoo's Print Journal,* 22 (3), pp. 2599 - 2607.

Karmakar, A. K. and Das, A. (2007), *State Fauna Series 14, Fauna of Mizoram* (*Fishes*), Zoological Survey of India, (Calcutta), pp.507-535.

Lundberg, J. G., J. P. Sullivan, R. Rodiles-Hernández, and D. A. Hendrickson. 2007. Discovery of African roots for the Mesoamerican Chiapas catfish, *Lacantunia enigmatica*, requires an ancient intercontinental passage. Proceedings of the Academy of Natural Sciences of Philadelphia 156: 39–53.

Mittermeir, R.A., P.R. Myers Ngil and C.G. Mittermeir. 2000. Hotspots. Earths biological richness and most endangered terrestrial ecoregion. *CEMEX and Conservation International*, pp. 430.

Shahnawaz, A. Venkateshwaralu, M. Somashekar. D.S. and Santosh, K. Fish diversity with relation to water quality of Bhadra River of Western Ghats (India). Environ Monit Assess (2010), 161: 83-91; Doi 10.1007/s10661-008-0729-0.

Sullivan, J. P., J. G. Lundberg, and M. Hardman. 2006. A phylogenetic analysis of the major groups of catfishes (Teleostei: Siluriformes) using rag1 and rag2 nuclear gene sequences. Molecular Phylogenetics and Evolution, 41: 636–662.

Talwar, P. K. and Jhingran, A. G. 1991, *Inland Fishes of India and adjacent countries* (Volume I and II). Oxford and IBH Publishing Co. Pvt. Ltd., New Delhi.

Teugels, G. G. 2003. State of the art of recent siluriform systematics. Pages 317–352 *in* G. Arratia, B. G. Kapoor, M. Chardon, and R. Diogo, editors. Catfishes, Volume 1. Science Publishers, Enfi eld, New Hampshire.

Thomas, K. R., George, M. J. and Babu, C. R. 2002, "Freshwater Fishes of Southern Kerala with Notes on the Distribution of Endemic and Endangered Species. *Journal of the Bombay Natural History Society,* 99 (1), pp. 47-53.

Winemiller, K. O. and Winemiller, L. C. K. 1996, "Comparative Ecology of Catfishes of the Upper Zambezi River Flood Plain", *Journal of Fish Biology,* 49, pp. 1043-1061.

Index

D

E

F

G

H

I

K

L

M

N

O

P

R

S

T

W

9 789354 614095

Printed at Repro India Ltd.